Teacher Edition

Eureka Math
Grade 6
Module 3

Special thanks go to the Gordan A. Cain Center and to the Department of Mathematics at Louisiana State University for their support in the development of *Eureka Math*.

Published by the non-profit Great Minds

Copyright © 2015 Great Minds. All rights reserved. No part of this work may be reproduced or used in any form or by any means — graphic, electronic, or mechanical, including photocopying or information storage and retrieval systems — without written permission from the copyright holder. "Great Minds" and "Eureka Math" are registered trademarks of Great Minds.

Printed in the U.S.A.
This book may be purchased from the publisher at eureka-math.org
10 9 8 7 6 5 4 3 2 1
ISBN 978-1-63255-387-4

A STORY OF RATIOS

Mathematics Curriculum

GRADE 6 • MODULE 3

Table of Contents[1]

Rational Numbers

Module Overview 2

Topic A: Understanding Positive and Negative Numbers on the Number Line (**6.NS.C.5**, **6.NS.C.6a**, **6.NS.C.6c**) 10

    Lesson 1: Positive and Negative Numbers on the Number Line—Opposite Direction and Value 12

    Lessons 2–3: Real-World Positive and Negative Numbers and Zero 20

    Lesson 4: The Opposite of a Number 37

    Lesson 5: The Opposite of a Number's Opposite 45

    Lesson 6: Rational Numbers on the Number Line 51

Topic B: Order and Absolute Value (**6.NS.C.6c**, **6.NS.C.7**) 61

    Lessons 7–8: Ordering Integers and Other Rational Numbers 63

    Lesson 9: Comparing Integers and Other Rational Numbers 79

    Lesson 10: Writing and Interpreting Inequality Statements Involving Rational Numbers 89

    Lesson 11: Absolute Value—Magnitude and Distance 101

    Lesson 12: The Relationship Between Absolute Value and Order 109

    Lesson 13: Statements of Order in the Real World 118

Mid-Module Assessment and Rubric 125
Topics A through B (assessment 1 day, return 1 day, remediation or further applications 1 day)

Topic C: Rational Numbers and the Coordinate Plane (**6.NS.C.6b**, **6.NS.C.6c**, **6.NS.C.8**) 140

    Lesson 14: Ordered Pairs 142

    Lesson 15: Locating Ordered Pairs on the Coordinate Plane 149

    Lesson 16: Symmetry in the Coordinate Plane 157

    Lesson 17: Drawing the Coordinate Plane and Points on the Plane 165

    Lesson 18: Distance on the Coordinate Plane 174

    Lesson 19: Problem Solving and the Coordinate Plane 179

End-of-Module Assessment and Rubric 185
Topics A through C (assessment 1 day, return 1 day, remediation or further applications 1 day)

---

[1] Each lesson is ONE day, and ONE day is considered a 45-minute period.

Grade 6 • Module 3
Rational Numbers

OVERVIEW

Students are familiar with the number line and determining the location of positive fractions, decimals, and whole numbers from previous grades. Students extend the number line (both horizontally and vertically) in Module 3 to include the opposites of whole numbers. The number line serves as a model to relate integers and other rational numbers to statements of order in real-world contexts. In this module's final topic, the number line model is extended to two-dimensions as students use the coordinate plane to model and solve real-world problems involving rational numbers.

Topic A focuses on the development of the number line in the opposite direction (to the left or below zero). Students use positive integers to locate negative integers, understanding that a number and its opposite are on opposite sides of zero and that both lie the same distance from zero. Students represent the opposite of a positive number as a negative number and vice versa. Students realize that zero is its own opposite and that the opposite of the opposite of a number is actually the number itself (**6.NS.C.6a**). They use positive and negative numbers to represent real-world quantities, such as -50 to represent a $50 debt or 50 to represent a $50 deposit into a savings account (**6.NS.C.5**). Topic A concludes with students furthering their understanding of signed numbers to include the rational numbers. Students recognize that finding the opposite of any rational number is the same as finding an integer's opposite (**6.NS.C.6c**) and that two rational numbers that lie on the same side of zero have the same sign, while those that lie on opposites sides of zero have opposite signs.

In Topic B, students apply their understanding of a rational number's position on the number line (**6.NS.C.6c**) to order rational numbers. Students understand that when using a conventional horizontal number line, the numbers increase as they move along the line to the right and decrease as they move to the left. They recognize that if a and b are rational numbers and $a < b$, then it must be true that $-a > -b$. Students compare rational numbers using inequality symbols and words to state the relationship between two or more rational numbers. They describe the relationship between rational numbers in real-world situations and with respect to numbers' positions on the number line (**6.NS.C.7a**, **6.NS.C.7b**). For instance, students explain that $-10°F$ is warmer than $-11°F$ because -10 is to the right (or above) -11 on a number line and write $-10°F > -11°F$. Students use the concept of absolute value and its notation to show a number's distance from zero on the number line and recognize that opposite numbers have the same absolute value (**6.NS.C.7c**). In a real-world scenario, students interpret absolute value as magnitude for a positive or negative quantity. They apply their understanding of order and absolute value to determine that, for instance, a checking account balance that is less than -25 dollars represents a debt of more than $25 (**6.NS.C.7d**).

In Topic C, students extend their understanding of the ordering of rational numbers in one dimension (on a number line) to the two-dimensional space of the coordinate plane. They construct the plane's vertical and horizontal axes, discovering the relationship between the four quadrants and the signs of the coordinates of points that lie in each quadrant (**6.NS.C.6b**, **6.NS.C.6c**). Students build upon their foundational understanding

A STORY OF RATIOS   Module Overview   6•3

from Grade 5 (**5.G.A.1**, **5.G.A.2**) of plotting points in the first quadrant and transition to locating points in all four quadrants. Students apply the concept of absolute value to find the distance between points located on vertical or horizontal lines and solve real-world problems related to distance, segments, and shapes (**6.NS.C.8**).

The 25-day module consists of 19 lessons; 6 days are reserved for administering the Mid- and End-of-Module Assessments, returning assessments, and remediating or providing further applications of the concepts. The Mid-Module Assessment follows Topic B, and the End-of-Module Assessment follows Topic C.

Focus Standards

Apply and extend previous understandings of numbers to the system of rational numbers.

6.NS.C.5   Understand that positive and negative numbers are used together to describe quantities having opposite directions or values (e.g., temperature above/below zero, elevation above/below sea level, credits/debits, positive/negative electric charge); use positive and negative numbers to represent quantities in real-world contexts, explaining the meaning of 0 in each situation.

6.NS.C.6   Understand a rational number as a point on the number line. Extend number line diagrams and coordinate axes familiar from previous grades to represent points on the line and in the plane with negative number coordinates.

   a. Recognize opposite signs of numbers as indicating locations on opposite sides of 0 on the number line; recognize that the opposite of the opposite of a number is the number itself, e.g., $-(-3) = 3$, and that 0 is its own opposite.

   b. Understand signs of numbers in ordered pairs as indicating locations in quadrants of the coordinate plane; recognize that when two ordered pairs differ only by signs, the locations of the points are related by reflections across one or both axes.

   c. Find and position integers and other rational numbers on a horizontal or vertical number line diagram; find and position pairs of integers and other rational numbers on a coordinate plane.

6.NS.C.7   Understand ordering and absolute value of rational numbers.

   a. Interpret statements of inequality as statements about the relative position of two numbers on a number line diagram. *For example, interpret $-3 > -7$ as a statement that -3 is located to the right of -7 on a number line oriented from left to right.*

   b. Write, interpret, and explain statements of order for rational numbers in real-world contexts. *For example, write $-3°C > -7°C$ to express the fact that $-3°C$ is warmer than $-7°C$.*

Module 3:   Rational Numbers   3

This work is derived from Eureka Math ™ and licensed by Great Minds. ©2015 Great Minds. eureka-math.org
G6-M3-TE-B3-1.3.0-07.2015

c. Understand the absolute value of a rational number as its distance from 0 on the number line; interpret absolute value as magnitude for a positive or negative quantity in a real-world situation. *For example, for an account balance of −30 dollars, write |−30| = 30 to describe the size of the debt in dollars.*

d. Distinguish comparisons of absolute value from statements about order. *For example, recognize that an account balance less than −30 dollars represents a debt greater than 30 dollars.*

6.NS.C.8 Solve real-world and mathematical problems by graphing points in all four quadrants of the coordinate plane. Include use of coordinates and absolute value to find distances between points with the same first coordinate or the same second coordinate.

Foundational Standards

Develop understanding of fractions as numbers.

3.NF.A.2 Understand a fraction as a number on the number line; represent fractions on a number line diagram.

a. Represent a fraction $1/b$ on a number line diagram by defining the interval from 0 to 1 as the whole and partitioning it into b equal parts. Recognize that each part has size $1/b$ and that the endpoint of the part based at 0 locates the number $1/b$ on the number line.

b. Represent a fraction a/b on a number line diagram by marking off a lengths $1/b$ from 0. Recognize that the resulting interval has size a/b and that its endpoint locates the number a/b on the number line.

Draw and identify lines and angles, and classify shapes by properties of their lines and angles.

4.G.A.3 Recognize a line of symmetry for a two-dimensional figure as a line across the figure such that the figure can be folded along the line into matching parts. Identify line-symmetric figures and draw lines of symmetry.

Graph points on the coordinate plane to solve real-world and mathematical problems.

5.G.A.1 Use a pair of perpendicular number lines, called axes, to define a coordinate system, with the intersection of the lines (the origin) arranged to coincide with the 0 on each line and a given point in the plane located by using an ordered pair of numbers, called its coordinates. Understand that the first number indicates how far to travel from the origin in the direction of one axis, and the second number indicates how far to travel in the direction of the second axis, with the convention that the names of the two axes and the coordinates correspond (e.g., x-axis and x-coordinate, y-axis and y-coordinate).

5.G.A.2	Represent real-world and mathematical problems by graphing points in the first quadrant of the coordinate plane, and interpret coordinate values of points in the context of the situation.

Focus Standards for Mathematical Practice

MP.2	**Reason abstractly and quantitatively.** Students read a word problem involving integers, draw a number line or coordinate plane model, and write about their conclusions. They understand the meaning of quantities as they relate to the real world. For instance, a loss of 14 yards in a football game can be represented by -14, and a distance of 25 feet below sea level is greater than a distance of 5 feet above sea level because $|-25| > |5|$. Students decontextualize word problems related to distance by creating number lines and coordinate plane models. In doing so, they count the number of units between endpoints and use the concept of absolute value to justify their answers. For instance, when given the coordinate $(2, 6)$, students determine that the point $(2, -6)$ would be the same distance from the x-axis but in the opposite direction because both points have the same x-coordinate and their y-coordinates (6 and -6) have the same absolute value.
MP.4	**Model with mathematics.** Students use vertical and horizontal number lines to visualize integers and better understand their connection to whole numbers. They divide number line intervals into sub-intervals of tenths to determine the correct placement of rational numbers. Students may represent a decimal as a fraction or a fraction as a decimal to better understand its relationship to other rational numbers to which it is being compared. To explain the meaning of a quantity in a real-life situation (involving elevation, temperature, or direction), students may draw a diagram and/or number line to illustrate the location of the quantity in relation to zero or an established level that represents zero in that situation.
MP.6	**Attend to precision.** In representing signed numbers on a number line or as a quantity, students pay close attention to the direction and sign of a number. They realize that a negative number must lie to the left of zero on a horizontal number line or below zero on a vertical number line. They recognize that the way they represent their answer depends on the phrasing of a question and context of a word problem. For instance, a question that asks a student "How many feet below sea level is the diver?" would require the answer to be stated as a positive number, whereas, a question that is phrased "Which integer would represent 40 feet below sea level?" would require the answer to be written as -40.
MP.7	**Look for and make use of structure.** Students understand the placement of negative numbers on a number line by observing the patterns that exist between negative and positive numbers with respect to zero. They recognize that two numbers are opposites if they are the same distance from zero and that zero is its own opposite. Students extend their understanding of the number line's structure to the coordinate plane to determine a point's location. They recognize the relationship between the signs of a point's coordinates and the quadrant in which the point lies.

Module 3: Rational Numbers

Terminology

New or Recently Introduced Terms

- **Absolute Value** (The *absolute value* of a number is the distance between the number and zero on the number line. For example, $|3| = 3$, $|-4| = 4$, etc.)
- **Integer** (An *integer* is a number that can either be represented as a whole number or as the opposite of a whole number. The *set of integers* is the infinite list of numbers: $\ldots, -3, -2, -1, 0, 1, 2, 3, \ldots$.)
- **Magnitude** (The *magnitude of a measurement* is the absolute value of the measure of the measurement. For example, the magnitude of the measurement $-25°F$ is 25.)
- **Negative Number** (A *negative number* is a number less than zero.)
- **Opposite** (Given a nonzero number a on the number line, the *opposite of a*, denoted $-a$, is the number on the number line such that (1) 0 is between a and $-a$, and (2) the distance between 0 and a is equal to the distance between 0 and $-a$. The opposite of 0 is 0.)
- **Positive Number** (A *positive number* is a number greater than zero.)
- **Quadrant (description)** (In the Cartesian plane, the two axes separate the plane into four regions called *quadrants*. The first quadrant consists of all the points whose x- and y-coordinates are both positive. The first, second, third, and fourth quadrants are identified counterclockwise around the origin in order starting with the first quadrant.)
- **Rational Number (description)** (A *rational number* is a number that can be represented as a fraction or the opposite of a fraction.)

Familiar Terms and Symbols[2]

- Coordinate Pair
- Coordinate Plane
- Fraction
- Line of Symmetry
- Ordered Pair
- Origin
- Quadrant
- Symmetry
- Whole Numbers
- x-Axis
- x-Coordinate
- y-Axis
- y-Coordinate

---

[2] These are terms and symbols students have seen previously.

Suggested Tools and Representations

- Horizontal and Vertical Number Lines
- Coordinate Plane

Sprints

Sprints are designed to develop fluency. They should be fun, adrenaline-rich activities that intentionally build energy and excitement. A fast pace is essential. During Sprint administration, teachers assume the role of athletic coaches. A rousing routine fuels students' motivation to do their personal best. Student recognition of increasing success is critical, and so every improvement is acknowledged. (See the Sprint Delivery Script for the suggested means of acknowledging and celebrating student success.)

One Sprint has two parts with closely related problems on each. Students complete the two parts of the Sprint in quick succession with the goal of improving on the second part, even if only by one more.

Sprints are not to be used for a grade. Thus, there is no need for students to write their names on the Sprints. The low-stakes nature of the exercise means that even students with allowances for extended time can participate. When a particular student finds the experience undesirable, it is recommended that the student be allowed to opt out and take the Sprint home. In this case, it is ideal if the student has a regular opportunity to express the desire to opt in.

With practice, the Sprint routine takes about 8 minutes.

Sprint Delivery Script

Gather the following: stopwatch, a copy of Sprint A for each student, a copy of Sprint B for each student, answers for Sprint A and Sprint B. The following delineates a script for delivery of a pair of Sprints.

This sprint covers: *topic.*

Do not look at the Sprint; keep it turned facedown on your desk.

There are xx problems on the Sprint. You will have 60 seconds. Do as many as you can. I do not expect any of you to finish.

On your mark, get set, GO.

60 seconds of silence.

STOP. Circle the last problem you completed.

I will read the answers. You say "YES" if your answer matches. Mark the ones you have wrong. Don't try to correct them.

Energetically, rapid-fire call the answers ONLY.

Stop reading answers after there are no more students answering, "Yes."

Fantastic! Count the number you have correct, and write it on the top of the page. This is your personal goal for Sprint B.

Raise your hand if you have one or more correct. Two or more, three or more, ...

Let us all applaud our runner-up, [insert name], with x correct. And let us applaud our winner, [insert name], with x correct.

You have a few minutes to finish up the page and get ready for the next Sprint.

Students are allowed to talk and ask for help; let this part last as long as most are working seriously.

Stop working. I will read the answers again so you can check your work. You say "YES" if your answer matches.

Energetically, rapid-fire call the answers ONLY.

Optionally, ask students to stand, and lead them in an energy-expanding exercise that also keeps the brain going. Examples are jumping jacks or arm circles, etc., while counting by 15's starting at 15, going up to 150 and back down to 0. You can follow this first exercise with a cool-down exercise of a similar nature, such as calf raises with counting by one-sixths $\left(\frac{1}{6}, \frac{1}{3}, \frac{1}{2}, \frac{2}{3}, \frac{5}{6}, 1, ...\right)$.

Hand out the second Sprint, and continue reading the script.

Keep the Sprint facedown on your desk.

There are xx problems on the Sprint. You will have 60 seconds. Do as many as you can. I do not expect any of you to finish.

On your mark, get set, GO.

60 seconds of silence.

STOP. Circle the last problem you completed.

I will read the answers. You say "YES" if your answer matches. Mark the ones you have wrong. Don't try to correct them.

Quickly read the answers ONLY.

Count the number you have correct, and write it on the top of the page.

Raise your hand if you have one or more correct. Two or more, three or more, ...

Let us all applaud our runner-up, [insert name], with x correct. And let us applaud our winner, [insert name], with x correct.

Write the amount by which your score improved at the top of the page.

Raise your hand if you have one or more correct. Two or more, three or more, ...

Let us all applaud our runner-up for most improved, [insert name]. And let us applaud our winner for most improved, [insert name].

You can take the Sprint home and finish it if you want.

Assessment Summary

Assessment Type	Administered	Format	Standards Addressed
Mid-Module Assessment Task	After Topic B	Constructed response with rubric	6.NS.C.5, 6.NS.C.6a, 6.NS.C.6c, 6.NS.C.7
End-of-Module Assessment Task	After Topic C	Constructed response with rubric	6.NS.C.5, 6.NS.C.6a, 6.NS.C.6c, 6.NS.C.7, 6.NS.C.8

A STORY OF RATIOS

Mathematics Curriculum

GRADE 6 • MODULE 3

Topic A

Understanding Positive and Negative Numbers on the Number Line

6.NS.C.5, 6.NS.C.6a, 6.NS.C.6c

Focus Standards:	6.NS.C.5	Understand that positive and negative numbers are used together to describe quantities having opposite directions or values (e.g., temperature above/below zero, elevation above/below sea level, credits/debits, positive/negative electric charge); use positive and negative numbers to represent quantities in real-world contexts, explaining the meaning of 0 in each situation.
	6.NS.C.6a 6.NS.C.6c	Understand a rational number as a point on the number line. Extend number line diagrams and coordinate axes familiar from previous grades to represent points on the line and in the plane with negative number coordinates. a. Recognize opposite signs of numbers as indicating locations on opposite sides of 0 on the number line; recognize that the opposite of the opposite of a number is the number itself, e.g., $-(-3) = 3$, and that 0 is its own opposite. c. Find and position integers and other rational numbers on a horizontal or vertical number line diagram; find and position pairs of integers and other rational numbers on a coordinate plane.
Instructional Days:	6	
Lesson 1:		Positive and Negative Numbers on the Number Line—Opposite Direction and Value (E)[1]
Lessons 2–3:		Real-World Positive and Negative Numbers and Zero (P, E)
Lesson 4:		The Opposite of a Number (P)
Lesson 5:		The Opposite of a Number's Opposite (P)
Lesson 6:		Rational Numbers on the Number Line (S)

---

[1]Lesson Structure Key: **P**-Problem Set Lesson, **M**-Modeling Cycle Lesson, **E**-Exploration Lesson, **S**-Socratic Lesson

A STORY OF RATIOS   Topic A   6•3

In Topic A, students apply their understanding of the ordering of whole numbers, positive fractions, and decimals to extend the number line in the opposite direction (**6.NS.C.6 stem**). In Lessons 1–3, students use positive integers to locate negative integers on the number line, moving in the opposite direction from zero, realizing that zero is its own opposite. They represent real-world situations with integers (**6.NS.C.5**) and understand the vocabulary and context related to opposite quantities (e.g., deposit/withdraw, elevation above/below sea level, debit/credit). Students use precise vocabulary to state, for instance, that -10 would describe an elevation that is 10 feet below sea level. In Lessons 4 and 5, students focus on locating the opposite of a number and the opposite of an opposite, using zero and the symmetry of the number line to build a conceptual understanding (**6.NS.C.6a**). In Lesson 6, students extend their understanding of integers to locate signed non-integer rational numbers on the number line (**6.NS.C.6c**), realizing that finding the opposite of any rational number is the same as finding an integer's opposite.

Topic A: Understanding Positive and Negative Numbers on the Number Line

Lesson 1: Positive and Negative Numbers on the Number Line—Opposite Direction and Value

Student Outcomes

- Students extend their understanding of the number line, which includes zero and numbers to the right or above zero that are greater than zero and numbers to the left or below zero that are less than zero.
- Students use positive integers to locate negative integers by moving in the opposite direction from zero.
- Students understand that the set of integers includes the set of positive whole numbers and their opposites, as well as zero. They also understand that zero is its own opposite.

Lesson Notes

Each student needs a compass to complete the Exploratory Challenge in this lesson.

Classwork

Opening Exercise (3 minutes): Number Line Review

Display two number lines (horizontal and vertical), each numbered 0–10. Allow students to discuss the following questions in cooperative learning groups of three or four students each.

Students should remain in the groups for the entire lesson.

Discuss the following:

- What is the starting position on both number lines?
  - *0 (zero)*
- What is the last whole number depicted on both number lines?
  - *10 (ten)*
- On a horizontal number line, do the numbers increase or decrease as you move farther to the right of zero?
  - *Increase*
- On a vertical number line, do the numbers increase or decrease as you move farther above zero?
  - *Increase*

Scaffolding:
- *Create two floor models for vertical and horizontal number lines using painter's tape for visual learners.*
- *Have a student model movement along each number line for kinesthetic learners.*
- *Use polling software for questions to gain immediate feedback while accessing prior knowledge.*

A STORY OF RATIOS   Lesson 1   6•3

Exploratory Challenge (10 minutes): Constructing the Number Line

The purpose of this exercise is to let students construct the number line (positive and negative numbers and zero) using a compass.

Have students draw a line, place a point on the line, and label it 0.

Have students use the compass to locate and label the next point 1, thus creating the scale. Students continue to locate other whole numbers to the right of zero using the same unit measure.

Using the same process, have students locate the opposites of the whole numbers. Have students label the first point to the left of zero -1.

Introduce to the class the definition of the opposite of a number.

Sample student work is shown below.

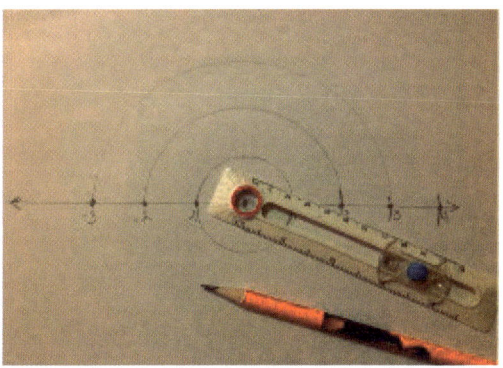

> Given a nonzero number, a, on a number line, the opposite of a, labeled $-a$, is the number such that
>
> - 0 is between a and $-a$.
> - The distance between 0 and a is equal to the distance between 0 and $-a$.
>
> The opposite of 0 is 0.

- The set of whole numbers and their opposites, including zero, are called **integers**. Zero is its own opposite. The number line diagram shows integers listed in order from least to greatest using equal spaces.

 Monitor student constructions, making sure students are paying close attention to the direction and sign of a number.

Example 1 (5 minutes): Negative Numbers on the Number Line

Students use their constructions to model the location of a number relative to zero by using a curved arrow starting at zero and pointing away from zero toward the number. Pose questions to students as a whole group, one question at a time.

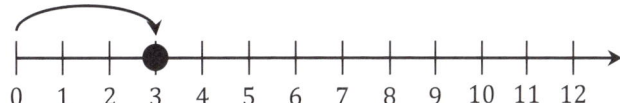

   Lesson 1:   Positive and Negative Numbers on the Number Line—Opposite Direction and Value   13

This work is derived from Eureka Math ™ and licensed by Great Minds. ©2015 Great Minds. eureka-math.org
G6-M3-TE-B3-1.3.0-07.2015

- Starting at 0, as I move to the right on a horizontal number line, the values get larger. These numbers are called *positive numbers* because they are greater than zero. Notice the curved arrow is pointing to the right to show a positive direction.
- How far is the number from zero on the number line?
  - 3 *units*
- If 0 was a mirror facing toward the arrow, what would be the direction of the arrow in the mirror?
  - *To the left*
- Would the numbers get larger or smaller as we move to the left of zero?
  - *Smaller*
- Starting at 0, as I move farther to the left of zero on a horizontal number line, the values get smaller. These numbers are called *negative numbers* because they are less than zero. Notice the curved arrow is pointing to the left to show a negative direction. The position of the point is now at negative 3, written as −3.

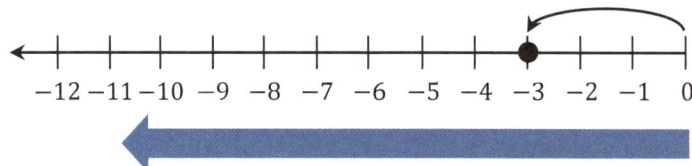

- Negative numbers are less than zero. As you move to the left on a horizontal number line, the values of the numbers decrease.

- What is the relationship between 3 and −3 on the number line?
  - *3 and −3 are located on opposite sides of zero. They are both the same distance from zero. 3 and −3 are called opposites.*
- As we look farther right on the number line, the values of the numbers increase. For example, $-1 < 0 < 1 < 2 < 3$.
- This is also true for a vertical number line. On a vertical number line, positive numbers are located above zero. As we look upward on a vertical number line, the values of the numbers increase. On a vertical number line, negative numbers are located below zero. As we look farther down on a vertical number line, the values of the numbers decrease.
- The set of whole numbers and their opposites, including zero, are called *integers*. Zero is its own opposite. A number line diagram shows integers listed in increasing order from left to right, or from bottom to top, using equal spaces. For example: −4, −3, −2, −1, 0, 1, 2, 3, 4.

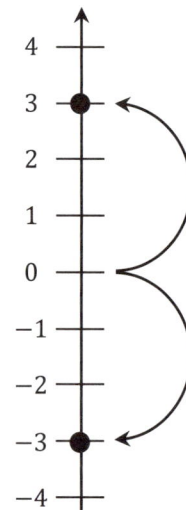

Allow students to discuss the example in their groups to solidify their understanding of positive and negative numbers.

Possible discussion questions:

- Where are negative numbers located on a horizontal number line?
  - *Negative numbers are located to the left of 0 on a horizontal number line.*

- Where are negative numbers located on a vertical number line?
  - *Negative numbers are located below 0 on a vertical number line.*
- What is the opposite of 2?
  - *−2*
- What is the opposite of 0?
  - *0*
- Describe the relationship between 10 and −10.
  - *10 and −10 are opposites because they are on opposite sides of 0 and are both 10 units from 0.*

Example 2 (5 minutes): Using Positive Integers to Locate Negative Integers on the Number Line

Have students establish elbow partners, and tell them to move their fingers along their number lines to answer the following set of questions. Students can discuss answers with their elbow partners. Circulate around the room, and listen to the student–partner discussions.

- Describe to your elbow partner how to find 4 on a number line. Describe how to find −4.
  - *To find 4, start at zero, and move right to 4. To find −4, start at zero, and move left to −4.*

> **Scaffolding:**
>
> As an extension activity, have students identify the *unit* differently on different number lines, and ask students to locate two whole numbers other than 1 and their opposites.

Model how the location of a positive integer can be used to locate a negative integer by moving in the opposite direction.

- Explain and show how to find 4 and the opposite of 4 on a number line.
  - *Start at zero, and move 4 units to the right to locate 4 on the number line. To locate −4, start at zero, and move 4 units to the left on the number line.*
- Where do you start when locating an integer on the number line?
  - *Always start at zero.*
- What do you notice about the curved arrows that represent the location of 4 and −4?
  - *They are the same distance but pointing in opposite directions.*

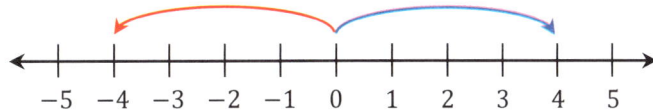

Exercises 1–5 (13 minutes)

Create and display two large number lines on the board, one horizontal and one vertical, each numbered −12 to 12, for the class. Distribute an index card with an integer from −12 to 12 on it to each group.[1] Students work in groups first, completing the exercises in their student materials. Conclude the exercise by having students locate and label their integers on the teacher's number lines.

---

[1] Depending on the class size, label enough index cards for ten groups (one card per group). Vary the numbers using positives and negatives, such as −5 through 5, including zero. If each group finds and locates the integer correctly, each group will have a card that is the opposite of another group's card.

Exercises

Complete the diagrams. Count by ones to label the number lines.

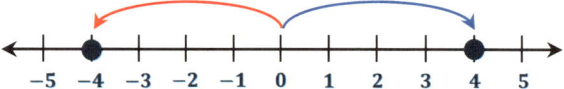

1. Plot your point on both number lines.

   Answers may vary.

2. Show and explain how to find the opposite of your number on both number lines.

   In this example, the number chosen was −4. So −4 is the first number plotted, and the opposite is 4.

   Horizontal Number Line: I found my point by starting at zero and counting four units to the left to end on −4. Then, to find the opposite of my number, I started on zero and counted to the right four units to end on 4.

   Vertical Number Line: I found my point by starting at zero and counting four units down to end on −4.

   I found the opposite of my number by starting at zero and counting four units up to end on 4.

3. Mark the opposite on both number lines.

   Answers may vary.

4. Choose a group representative to place the opposite number on the class number lines.

5. Which group had the opposite of the number on your index card?

   Answers may vary. Jackie's group had the opposite of the number on my index card. They had 4.

Closing (2 minutes)

- Give an example of two opposite numbers, and describe their locations first on a horizontal and then on a vertical number line.
    - *For example, 6 and −6 are the same distance from zero but on opposite sides. Positive 6 is located 6 units to the right of zero on a horizontal number line and 6 units above zero on a vertical number line. Negative 6 is located 6 units to the left of zero on a horizontal number line and 6 units below zero on a vertical number line.*

Exit Ticket (7 minutes)

Name _____ Date _____

Lesson 1: Positive and Negative Numbers on the Number Line— Opposite Direction and Value

Exit Ticket

1. If zero lies between a and d, give one set of possible values for a, b, c, and d.

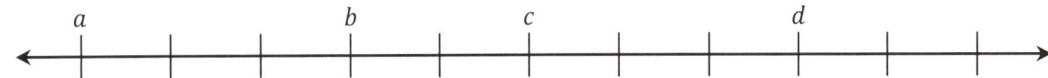

2. Below is a list of numbers in order from least to greatest. Use what you know about the number line to complete the list of numbers by filling in the blanks with the missing integers.

   $-6, -5,$ _____, $-3, -2, -1,$ _____, $1, 2,$ _____, $4,$ _____, 6

3. Complete the number line scale. Explain and show how to find 2 and the opposite of 2 on a number line.

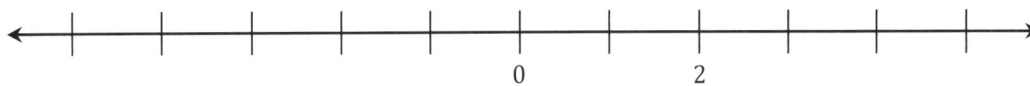

A STORY OF RATIOS   Lesson 1   6•3

Exit Ticket Sample Solutions

1. If zero lies between a and d, give one set of possible values for $a, b, c,$ and d.

   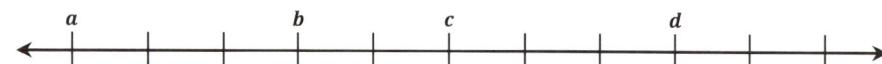

   Answers will vary. One possible answer is a: -4; b: -1; c: 1; d: 4.

2. Below is a list of numbers in order from least to greatest. Use what you know about the number line to complete the list of numbers by filling in the blanks with the missing integers.

   $-6, -5, \underline{\ -4\ }, -3, -2, -1, \underline{\ 0\ }, 1, 2, \underline{\ 3\ }, 4, \underline{\ 5\ }, 6$

3. Complete the number line scale. Explain and show how to find 2 and the opposite of 2 on a number line.

   I would start at zero and move 2 units to the left to locate the number -2 on the number line. So, to locate 2, I would start at zero and move 2 units to the right (the opposite direction).

Problem Set Sample Solutions

1. Draw a number line, and create a scale for the number line in order to plot the points -2, 4, and 6.

   a. Graph each point and its opposite on the number line.

   b. Explain how you found the opposite of each point.

   To graph each point, I started at zero and moved right or left based on the sign and number (to the right for a positive number and to the left for a negative number). To graph the opposites, I started at zero, but this time I moved in the opposite direction the same number of times.

2. Carlos uses a vertical number line to graph the points -4, -2, 3, and 4. He notices that -4 is closer to zero than -2. He is not sure about his diagram. Use what you know about a vertical number line to determine if Carlos made a mistake or not. Support your explanation with a number line diagram.

   Carlos made a mistake because -4 is less than -2, so it should be farther down the number line. Starting at zero, negative numbers decrease as we look farther below zero. So, -2 lies before -4 on a number line since -2 is 2 units below zero and -4 is 4 units below zero.

Lesson 1:   Positive and Negative Numbers on the Number Line—Opposite Direction and Value

3. Create a scale in order to graph the numbers −12 through 12 on a number line. What does each tick mark represent?

   Each tick mark represents 1 unit.

4. Choose an integer between −5 and −10. Label it R on the number line created in Problem 3, and complete the following tasks.

   Answers may vary. Refer to the number line above for sample student work. −6, −7, −8, or −9

   a. What is the opposite of R? Label it Q.

      Answers will vary. 6

   b. State a positive integer greater than Q. Label it T.

      Answers will vary. 11

   c. State a negative integer greater than R. Label it S.

      Answers will vary. −3

   d. State a negative integer less than R. Label it U.

      Answers will vary. −9

   e. State an integer between R and Q. Label it V.

      Answers will vary. 2

5. Will the opposite of a positive number *always, sometimes,* or *never* be a positive number? Explain your reasoning.

   The opposite of a positive number will never be a positive number. For two nonzero numbers to be opposites, zero has to be in between both numbers, and the distance from zero to one number has to equal the distance between zero and the other number.

6. Will the opposite of zero *always, sometimes,* or *never* be zero? Explain your reasoning.

   The opposite of zero will always be zero because zero is its own opposite.

7. Will the opposite of a number *always, sometimes,* or *never* be greater than the number itself? Explain your reasoning. Provide an example to support your reasoning.

   The opposite of a number will sometimes be greater than the number itself because it depends on the given number. For example, if the number given is −6, then the opposite is 6, which is greater than −6. If the number given is 5, then the opposite is −5, which is not greater than 5. If the number given is 0, then the opposite is 0, which is never greater than itself.

Lesson 1: Positive and Negative Numbers on the Number Line—Opposite Direction and Value

Lesson 2: Real-World Positive and Negative Numbers and Zero

Student Outcomes

- Students use positive and negative numbers to indicate a change (gain or loss) in elevation with a fixed reference point, temperature, and the balance in a bank account.
- Students use vocabulary precisely when describing and representing situations involving integers; for example, an elevation of -10 feet is the same as 10 feet below the fixed reference point.
- Students choose an appropriate scale for the number line when given a set of positive and negative numbers to graph.

Classwork

Opening Exercise (5 minutes)

Display a number line without a scale labeled. Pose the following questions to the whole group, and allow students three minutes to discuss their responses in pairs. Record feedback by labeling and relabeling the number line based on different responses.

Scaffolding:

For kinesthetic learners, provide students with white boards and markers to create their number lines. Ask them to hold up their boards, and select a few students to explain their diagrams to the class.

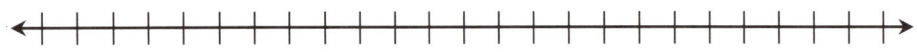

Discuss the following:

MP.4
- Explain how you would show 150 on a number line.
  - *I would start at zero and move to the right 150 units.*
- What strategy would you use to number the number line in order to show 150?
  - *I would locate (place) zero as far to the left as possible and use a scale of 10. I could also label the first tick mark 140 and count by ones.*
- If you want to have zero and 150 on the given number line, what scales would work well (what should you count by)?
  - *I could count by fives, tens, or twenty-fives.*

Common Misconceptions

Explain to students how to choose an appropriate scale. Pay careful attention to student graphs, and address common misconceptions, such as:

- Unequal intervals—Intervals should be equal from one mark to the next. This usually happens when students stop skip-counting in order to make the numbers fit on the diagram (e.g., 5, 10, 15, 20, 50, 100, 150).

- Miscounting—This is usually the result of students rushing and not paying attention to details. Students should always check their scales for accuracy before plotting points.

- Always starting at zero—The problem should determine the appropriate start and end point for a number line. Help struggling students by counting the number of tick marks (lines) first in order to determine a starting point.

- Not using the entire number line diagram—Spacing should be evenly distributed throughout a number line. This usually happens when students are counting by a value that is too large (e.g., counting by tens instead of twos).

Example 1 (10 minutes): Take It to the Bank

The purpose of this example is for students to understand how negative and positive numbers can be used to represent real-world situations involving money. Students are introduced to basic financial vocabulary—deposit, credit (credited), debit (debited), withdrawal, and change (gain or loss) throughout the example. The teacher should access prior knowledge by having students independently complete the first two columns of the KWL graphic organizer in their student materials. Monitor student responses, and select a few students to share out loud.

Example 1: Take It to the Bank

Read Example 1 silently. In the first column, write down any words and definitions you know. In the second column, write down any words you do not know.

For Tim's 13th birthday, he received $150 in cash from his mom. His dad took him to the bank to open a savings account. Tim gave the cash to the banker to deposit into the account. The banker credited Tim's new account $150 and gave Tim a receipt. One week later, Tim deposited another $25 that he had earned as allowance. The next month, Tim's dad gave him permission to withdraw $35 to buy a new video game. Tim's dad explained that the bank would charge a $5 fee for each withdrawal from the savings account and that each withdrawal and charge results in a debit to the account.

Words I Already Know:	Words I Want to Know:	Words I Learned:
Bank account—place where you put your money	Credited	
Receipt—ticket they give you to show how much you spent	Debit	
	Fee	
Allowance—money for chores	Deposit	
Charge—something you pay	Withdraw	

In the third column, write down any new words and definitions that you learn during the discussion.

Exercises 1–2 (7 minutes)

These exercises ask students to number the events of the story problem in order to show how each action can be represented by an integer and modeled on a number line. Record the events in the diagram below.

- Complete the first exercise, and then wait for further instruction.

A STORY OF RATIOS   Lesson 2 6•3

> **Exercises 1–2**
>
> 1. Read Example 1 again. With your partner, number the events in the story problem. Write the number above each sentence to show the order of the events.
>
> ①   ②
> For Tim's 13th birthday, he received $150 in cash from his mom. His dad took him to the bank to open a savings account.
>
>   ③   ④
> Tim gave the cash to the banker to deposit into the account. The banker credited Tim's new account $150 and gave Tim
>   ⑤
> a receipt. One week later, Tim deposited another $25 that he had earned as allowance. The next month, Tim's dad gave
>   ⑥
> him permission to withdraw $35 to buy a new video game. Tim's dad explained that the bank would charge a $5 fee for
>   ⑦
> each withdrawal from the savings account and that each withdrawal and charge results in a debit to the account.

After students complete Exercise 1, precisely define vocabulary to describe each situation as an integer, and model the integer on a number line. Pose the following questions throughout the exercise. Point out that zero represents the balance before each transaction in the story problem.

Discuss the following:

- Tim receives $150 for his birthday. Do you think this will be a positive or negative number for Tim's money? Explain.
    - *Positive; $150 is a gain for Tim's money. Positive numbers are greater than 0.*
- How much money is in the account when Tim opened it? What does this number represent in this situation?
    - *The account has $0 in it because Tim had not put in or taken out any money. Zero represents the starting account balance.*
- The $150 that Tim gives the banker is called a *deposit*. A deposit is the act of putting money into a bank account. To show the amount of money in Tim's savings account, would this deposit be located to the left or right of zero on the horizontal number line?
    - *This deposit is located to the right of zero because it increases the amount of money in the savings account.*
- The bank credited the account $150. A *credit* is when money is deposited into an account. The account increases in value. How would you represent a credit of $150 as an integer? Explain.
    - *Since a credit is a deposit and deposits are written as positive numbers, then positive 150 represents a credit of $150.*
- Tim makes another deposit of $25. Would this be a positive or negative number for Tim's savings account, and how would you show it on a horizontal number line?
    - *A deposit increases the amount of money in the savings account, so 25 is positive. I would place the point 25 units to the right of zero.*
- The bank creates a debit of $5 for any withdrawal. What do you think the word *debit* means in this situation?
    - *A debit sounds like the opposite of a credit. It might be something taken away. Taking money out of the savings account is the opposite of putting money in.*

22   Lesson 2:   Real-World Positive and Negative Numbers and Zero

- A debit means money paid out of an account. It is the opposite of a credit. Are debits represented as positive or negative numbers on the horizontal number line for the amount of money in a savings account?
  - *A debit is represented as a negative number to the left of zero on a number line because debits are the opposite of credits, which are positive numbers.*
- The bank charges a $5 service fee for any withdrawal from a savings account. A *charge*, also called a *fee*, is the amount of money a person has to pay for something. Can you name a situation where you would have to pay a charge?
  - *I would have to pay a charge at an amusement park, a concert, a basketball game, or a doctor's office.*
- How would you represent a charge of $5 for Tim's savings account on the horizontal number line?
  - *A charge of $5 would be −5 because money is being taken out of the account. I would find positive five on the number line by starting at 0 and moving 5 units to the right. Then, I would count 5 units going left of zero to end at −5.*
- Tim withdrew $35 from his account. Based on the story problem, what is the meaning of the term *withdraw*?
  - *Since Tim wanted to buy something, he took money out of the account. I think withdraw means to take money out of an account.*
- To withdraw money is to take money out of an account. How would you represent the $35 for the video game as an integer for Tim's savings account?
  - *The money was taken out of Tim's account; it would be represented as −35.*

2. Write each individual description below as an integer. Model the integer on the number line using an appropriate scale.

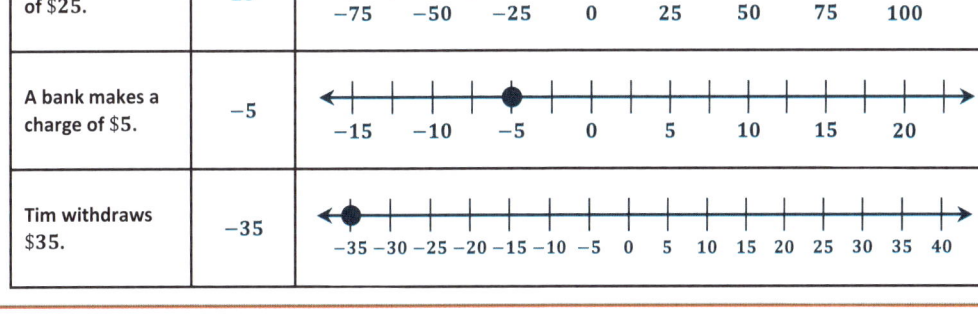

Lesson 2: Real-World Positive and Negative Numbers and Zero

A STORY OF RATIOS   Lesson 2 6•3

Example 2 (7 minutes): How Hot, How Cold?

This example gives students practice reading thermometers in both Fahrenheit and Celsius scales. Students write temperatures as integers and describe how temperature could be modeled on a vertical number line.

> **Example 2: How Hot, How Cold?**
>
> Temperature is commonly measured using one of two scales, Celsius or Fahrenheit. In the United States, the Fahrenheit system continues to be the accepted standard for nonscientific use. All other countries have adopted Celsius as the primary scale in use. The thermometer shows how both scales are related.
>
> a. The boiling point of water is 100°C. Where is 100 degrees Celsius located on the thermometer to the right?
>
>   *It is not shown because the greatest temperature shown in Celsius is 50°C.*
>
> b. On a vertical number line, describe the position of the integer that represents 100°C.
>
>   *The integer is 100, and it would be located 100 units above zero on the Celsius side of the scale.*
>
>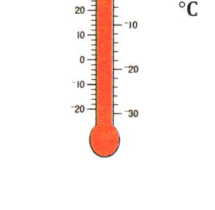
>
> c. Write each temperature as an integer.
>   i. The temperature shown on the thermometer in degrees Fahrenheit:
>
>   100
>
>   ii. The temperature shown on the thermometer in degrees Celsius:
>
>   38
>
>   iii. The freezing point of water in degrees Celsius:
>
>   0
>
> d. If someone tells you your body temperature is 98.6°, what scale is being used? How do you know?
>
>   *Since water boils at 100°C, they must be using the Fahrenheit scale.*
>
> e. Does the temperature 0 degrees mean the same thing on both scales?
>
>   *No. 0°C corresponds to 32°F, and 0°F corresponds to approximately −18°C.*

Address the common misconception on how to describe negative temperatures. −10°C can be read as "negative ten degrees Celsius." It can also be read as "ten degrees below zero." However, it should not be read as "negative ten degrees below zero."

Temperatures that are above zero can be stated as their numerical value. For example, describing a fever of 102°F can be simply stated as "one hundred two degrees."

> *Scaffolding:*
>
> Provide kinesthetic and visual learners with a thermometer to reinforce scales.

A STORY OF RATIOS   Lesson 2   6•3

Exercises 3–5 (7 minutes)

The following problems provide students additional practice with real-world positive and negative numbers and zero. Give students time to share responses to the whole group.

Exercises 3–5

3. Write each word under the appropriate column, "Positive Number" or "Negative Number."

   Gain   Loss   Deposit   Credit   Debit   Charge   Below Zero   Withdraw   Owe   Receive

Positive Number	Negative Number
Gain	Loss
Deposit	Debit
Credit	Charge
Receive	Below zero
	Withdraw
	Owe

4. Write an integer to represent each of the following situations:

   a. A company loses $345,000 in 2011. $-345,000$

   b. You earned $25 for dog sitting. 25

   c. Jacob owes his dad $5. -5

   d. The temperature at the sun's surface is about $5,500°C$ $5\ 500$

   e. The temperature outside is 4 degrees below zero. -4

   f. A football player lost 10 yards when he was tackled. -10

5. Describe a situation that can be modeled by the integer -15. Explain what zero represents in the situation.

   Answers will vary. I owe my best friend $15. In this situation, 0 represents my owing nothing to my best friend.

Closing (2 minutes)

- How did we represent debit and credit on a number line?
  - *A debit is represented as a negative number that is located to the left of (or below) zero. A credit is represented as a positive number that is located to the right of (or above) zero.*
- Can a temperature of -9 degrees be described as "Negative nine degrees below zero?" Why or why not?
  - *No, because "below zero" already means that the temperature is negative.*

Exit Ticket (7 minutes)

Lesson 2:   Real-World Positive and Negative Numbers and Zero   25

Name _____   Date _____

Lesson 2: Real-World Positive and Negative Numbers and Zero

Exit Ticket

1. Write a story problem that includes both integers −8 and 12.

2. What does zero represent in your story problem?

3. Choose an appropriate scale to graph both integers on the vertical number line. Label the scale.

4. Graph both points on the vertical number line.

A STORY OF RATIOS   Lesson 2   6•3

Exit Ticket Sample Solutions

1. Write a story problem that includes both integers −8 and 12.

   Answers may vary. One boxer gains 12 pounds of muscle to train for a fight. Another boxer loses 8 pounds of fat.

2. What does zero represent in your story problem?

   Zero represents no change in the boxer's weight.

3. Choose an appropriate scale to graph both integers on the vertical number line. Label the scale.

   I chose a scale of 1.

4. Graph both points on the vertical number line.

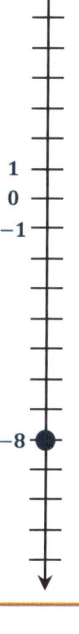

Problem Set Sample Solutions

1. Express each situation as an integer in the space provided.

a.	A gain of 56 points in a game	56
b.	A fee charged of $2	−2
c.	A temperature of 32 degrees below zero	−32
d.	A 56-yard loss in a football game	−56
e.	The freezing point of water in degrees Celsius	0
f.	A $12,500 deposit	12,500

Lesson 2:   Real-World Positive and Negative Numbers and Zero   27

For Problems 2–5, use the thermometer to the right.

2. Each sentence is stated *incorrectly*. Rewrite the sentence to correctly describe each situation.

   a. The temperature is -10 degrees Fahrenheit below zero.

   Correct: *The temperature is $-10°$F.*

   OR

   The temperature is 10 degrees below zero Fahrenheit.

   b. The temperature is -22 degrees Celsius below zero.

   Correct: *The temperature is $-22°$C.*

   OR

   The temperature is 22 degrees below zero Celsius.

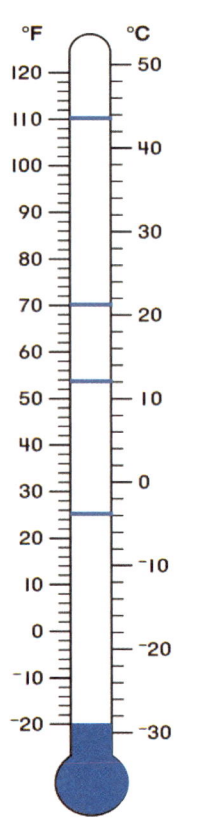

3. Mark the integer on the thermometer that corresponds to the temperature given.
   a. $70°$F
   b. $12°$C
   c. $110°$F
   d. $-4°$C

4. The boiling point of water is $212°$F. Can this thermometer be used to record the temperature of a boiling pot of water? Explain.

   No, it cannot because the highest temperature in Fahrenheit on this thermometer is $120°$.

5. Kaylon shaded the thermometer to represent a temperature of 20 degrees below zero Celsius as shown in the diagram. Is she correct? Why or why not? If necessary, describe how you would fix Kaylon's shading.

   She is incorrect because she shaded a temperature of $-20°$F. I would fix this by marking a line segment at $-20°$C and shade up to that line.

Lesson 2: Real-World Positive and Negative Numbers and Zero

Lesson 3: Real-World Positive and Negative Numbers and Zero

Student Outcomes

- Students use positive and negative numbers to indicate a change (gain or loss) in elevation with a fixed reference point, temperature, and the balance in a bank account.
- Students use vocabulary precisely when describing and representing situations involving integers; for instance, an elevation of -10 feet is the same as 10 feet below the fixed reference point.
- Students choose an appropriate scale for the number line when given a set of positive and negative numbers to graph.

Classwork

Example 1 (10 minutes): A Look at Sea Level

The purpose of this example is for students to understand how negative and positive numbers can be used to represent real-world situations involving elevation. Read the example aloud.

> **Example 1: A Look at Sea Level**
>
> The picture below shows three different people participating in activities at three different elevations. With a partner, discuss what you see. What do you think the word *elevation* means in this situation?

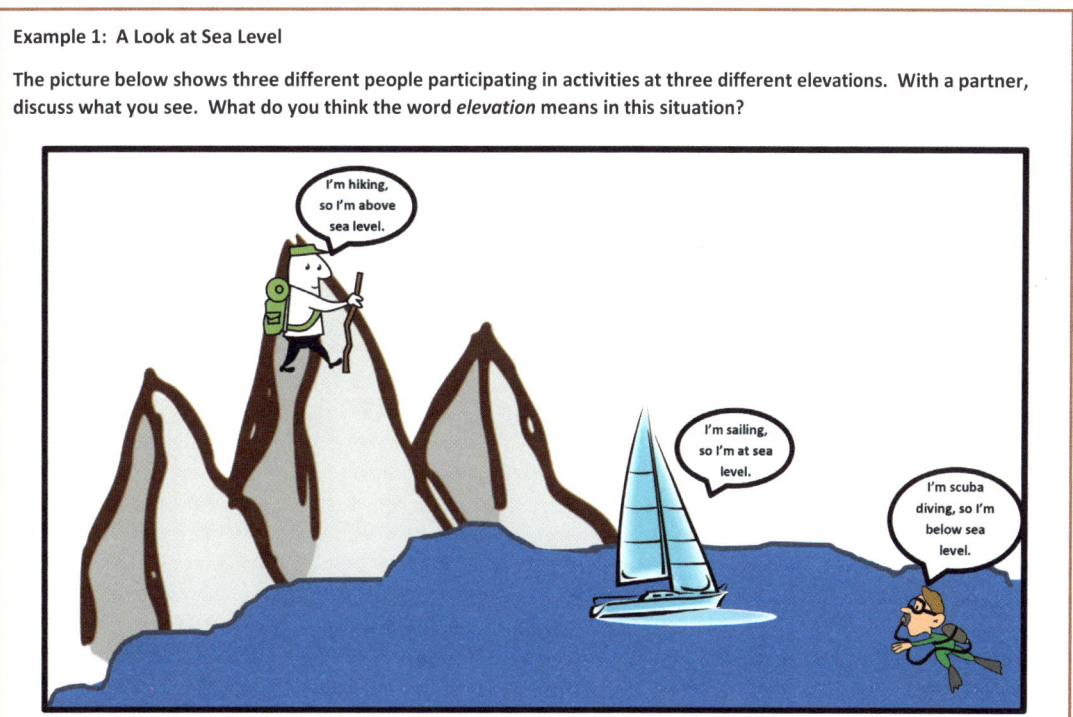

Lesson 3: Real-World Positive and Negative Numbers and Zero

A STORY OF RATIOS　　　　　　　　　　　　　　　　　　　　　　　　Lesson 3   6•3

Pose questions to the class, and define elevation. Students gain additional practice with elevation by completing Exercise 1 independently.

Possible discussion questions:

- Looking at the picture, if you were to draw a vertical number line to model elevation, which person's elevation do you think would be at zero? Explain.
  - *Sea level should represent an elevation of zero. So, the person sailing would be at zero because he is sailing on the surface of the water, which is neither above nor below the surface. On a number line, zero is the point or number separating positive and negative numbers.*
- On the same vertical number line, which person's elevation would be represented above zero?
  - *The elevation of the person hiking would be above zero because she is moving higher above the water. On a vertical number line, this is represented by a positive value above zero because she is above the surface.*
- On the same vertical number line, which person's elevation do you think would be below zero?
  - *The elevation of the person scuba diving would be below zero because he is swimming below the surface of the water. On a vertical number line, this is represented by a negative value below zero because he is below the surface.*
- What does zero represent in this situation?
  - *Zero represents the top of the water (the water's surface).*
- In this example, which numbers correspond to elevations above sea level?
  - *Above sea level means to be above zero, which are positive numbers.*
- In this example, which numbers correspond to elevations below sea level?
  - *Below sea level means to be below zero, which are negative numbers.*
- On a number line, what does it mean to be at sea level?
  - *To be at zero means to be at sea level.*
- Elevation is the height of a person, place, or thing above or below a certain reference point. In this case, what is the reference point?
  - *The reference point is sea level.*

Exercises 1–2 (5 minutes)

> **Exercises 1–3**
>
> Refer back to Example 1. Use the following information to answer the questions.
>
> - The scuba diver is 30 feet below sea level.
> - The sailor is at sea level.
> - The hiker is 2 miles (10,560 feet) above sea level.
>
> 1. Write an integer to represent each situation.
>
>   **Scuba Diver:** -30
>
>   **Sailor:** 0
>
>   **Hiker:** 2 *(to represent the elevation in miles) or* $10,560$ *(to represent the elevation in feet)*

30　　　Lesson 3:　　Real-World Positive and Negative Numbers and Zero

A STORY OF RATIOS   Lesson 3 6•3

2. Use an appropriate scale to graph each of the following situations on the number line to the right. Also, write an integer to represent both situations.

   a. A hiker is 15 feet above sea level.

      15

   b. A diver is 20 feet below sea level.

      −20

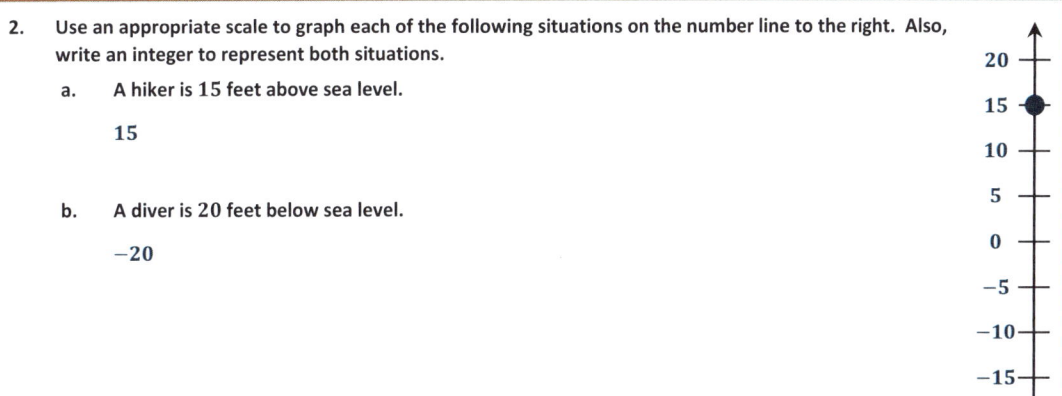

Students should identify common misconceptions of how to represent an answer based on the phrasing of a question. Students practice this skill in Exercise 3.

MP.6

- How many feet below sea level is the diver?
  - *Students should answer using a positive number, such as 70 feet, because "below" already indicates that the number is negative.*
- Which integer would represent 50 feet below sea level?
  - *Students should answer by saying "−50" and not "−50 below sea level."*

3. For each statement, there are two related statements: (i) and (ii). Determine which related statement ((i) or (ii)) is expressed correctly, and circle it. Then, correct the other related statement so that both parts, (i) and (ii), are stated correctly.

   a. A submarine is submerged 800 feet below sea level.
      i. The depth of the submarine is −800 feet below sea level.

      The depth of the submarine is 800 feet below sea level.

      ii. 800 feet below sea level can be represented by the integer −800. (circled)

   b. The elevation of a coral reef with respect to sea level is given as −150 feet.
      i. The coral reef is 150 feet below sea level. (circled)

      ii. The depth of the coral reef is −150 feet below sea level.

      The depth of the coral reef is 150 feet below sea level.

Lesson 3: Real-World Positive and Negative Numbers and Zero   31

Lesson 3

Exploratory Challenge (20 minutes)

Materials:

- Copies (one per student) of the Exploratory Challenge Station Record Sheet (see attached record sheet.)
- Sheets of loose-leaf paper (one per group) for the answer key for their posters
- Rulers or meter stick or yardstick (one per group)
- Construction paper or wall-sized grid paper (one sheet for each group)
- Markers (one set or a few for each group)

Students work in groups of three to four to create their own real-world situations involving money, temperature, elevation, and other real-world scenarios. Give each group a sheet of wall-sized grid paper (or construction paper) numbered one to five, markers, and a ruler. Using these materials, each group presents its situation on the paper by including the components in the bulleted list below. Allow students 10 minutes to create their posters, and hang them on a wall in the room.

> *Scaffolding:*
> Allow groups to present their posters to the class, and the class answers the questions during the presentation.

- Title (e.g., Sea Level, Temperature)
- A written situation based on the title (using at least two points)
- A blank vertical number line
- Picture (optional if time permits)
- Answer key (on a separate sheet of paper stapled to the top back right corner)

Groups rotate every few minutes to complete the three tasks on the Exploratory Challenge Station Record Sheet while viewing each poster.

- Write the integers for each situation.
- Determine the appropriate scale to graph the points.
- Graph the points on the number line.

Closing (3 minutes)

- How did we record measures of elevation on a number line?
  - *Elevations above sea level are positive numbers, and they are above zero. Elevations below sea level are negative numbers, and they are below zero.*
- Is "−90 feet below sea level" an appropriate answer to a question? Why or why not?
  - *No. You do not need the negative sign to write 90 feet below zero because the word "below" in this case means a negative number.*

Exit Ticket (7 minutes)

32   Lesson 3:   Real-World Positive and Negative Numbers and Zero

Lesson 3: Real-World Positive and Negative Numbers and Zero

Exit Ticket

1. Write a story problem using sea level that includes both integers −110 and 120.

2. What does zero represent in your story problem?

3. Choose an appropriate scale to graph both integers on the vertical number line.

4. Graph and label both points on the vertical number line.

A STORY OF RATIOS — Lesson 3 6•3

Exit Ticket Sample Solutions

1. Write a story problem using sea level that includes both integers −110 and 120.

   Answers may vary. On the beach, a man's kite flies at 120 feet above sea level, which is indicated by the water's surface. In the ocean, a white shark swims at 110 feet below the water's surface.

2. What does zero represent in your story problem?

   Zero represents the water's surface level, or sea level.

3. Choose and label an appropriate scale to graph both integers on the vertical number line.

   I chose a scale of 10.

4. Graph and label both points on the vertical number line.

Problem Set Sample Solutions

1. Write an integer to match the following descriptions.

a.	A debit of $40	−40
b.	A deposit of $225	225
c.	14,000 feet above sea level	14,000
d.	A temperature increase of 40°F	40
e.	A withdrawal of $225	−225
f.	14,000 feet below sea level	−14,000

For Problems 2–4, read each statement about a real-world situation and the two related statements in parts (a) and (b) carefully. Circle the correct way to describe each real-world situation; *possible answers include either (a), (b), or both (a) and (b)*.

2. A whale is 600 feet below the surface of the ocean.

   a. (The depth of the whale is 600 feet from the ocean's surface.) ← circled

   b. The whale is −600 feet below the surface of the ocean.

Lesson 3: Real-World Positive and Negative Numbers and Zero

3. The elevation of the bottom of an iceberg with respect to sea level is given as −125 feet.

   a. The iceberg is 125 feet above sea level.

   (b.) The iceberg is 125 feet below sea level.

4. Alex's body temperature decreased by 2°F.

   (a.) Alex's body temperature dropped 2°F.

   (b.) The integer −2 represents the change in Alex's body temperature in degrees Fahrenheit.

5. A credit of $35 and a debit of $40 are applied to your bank account.

   a. What is an appropriate scale to graph a credit of $35 and a debit of $40? Explain your reasoning.

   Answers will vary. I would count by 5's because both numbers are multiples of 5.

   b. What integer represents "a credit of $35" if zero represents the original balance? Explain.

   35; a credit is greater than zero, and numbers greater than zero are positive numbers.

   c. What integer describes "a debit of $40" if zero represents the original balance? Explain.

   −40; a debit is less than zero, and numbers less than zero are negative numbers.

   d. Based on your scale, describe the location of both integers on the number line.

   If the scale is multiples of 5, then 35 would be 7 units to the right of (or above) zero, and −40 would be 8 units to the left of (or below) zero.

   e. What does zero represent in this situation?

   Zero represents no change being made to the account balance. In other words, no amount is either subtracted or added to the account.

Lesson 3: Real-World Positive and Negative Numbers and Zero

A STORY OF RATIOS   Lesson 3   6•3

Name _____ Date _____

Exploratory Challenge Station Record Sheet

Poster # _____

Integers: _____

Number Line Scale: _____

Poster # _____

Integers: _____

Number Line Scale: _____

Poster # _____

Integers: _____

Number Line Scale: _____

Poster # _____

Integers: _____

Number Line Scale: _____

Poster # _____

Integers: _____

Number Line Scale: _____

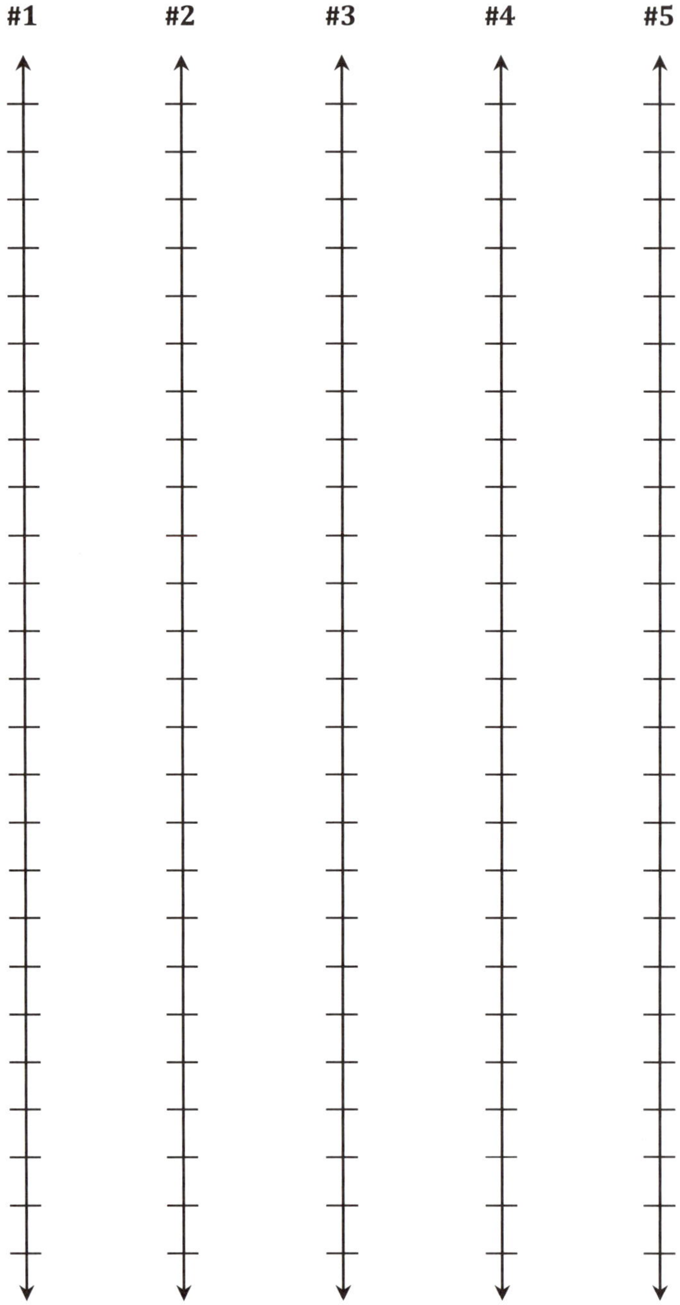

Lesson 3: Real-World Positive and Negative Numbers and Zero

Lesson 4: The Opposite of a Number

Student Outcomes

- Students understand that each nonzero integer, a, has an opposite, denoted $-a$, and that $-a$ and a are opposites if they are on opposite sides of zero and are the same distance from zero on the number line.
- Students recognize the number zero is its own opposite.
- Students understand that since all counting numbers are positive, it is not necessary to indicate such with a plus sign.

Lesson Notes

In this lesson, students practice graphing points on the number line. In particular, students determine the appropriate scale given a set of opposites in real-world situations. Students pay careful attention to the meaning of zero in problem situations and how opposites are related in the context of a given situation. Create a floor model of a number line prior to the lesson.

Classwork

Opening (5 minutes): What Is the Relationship?

Students work in pairs to determine the relationships between sets of words with or without pictures. Display the task to the whole group.

- Find the relationship between the sets of words.

Fast → Slow	Rough → Smooth	Open → Close	Fiction → Nonfiction
Light → Dark	Empty → Full	Accept → Refuse	Shallow → Deep
Dirty → Clean	Apart → Together	Question → Answer	Ancient → Modern
Alike → Different	All → None	Dangerous → Safe	Correct → Incorrect
Defeat → Victory	Easy → Hard	Future → Past	Break → Fix
Inside → Outside	Up → Down	Wet → Dry	Entrance → Exit

  □ The words are opposites of each other.

- Once you have determined the relationship, create your own examples, including a math example.

  □ Left → Right
  □ Cold → Hot
  □ 5 → −5

Scaffolding:
- Differentiate levels by providing groups with a set of 8–10 preselected words cut out individually on card stock. Use more challenging vocabulary words for advanced learners, and provide pictures with words for English language learners or inclusion students.
- Ask language arts and science teachers for input to provide more variation in vocabulary.

Lesson 4

Exercise 1 (10 minutes): Walk the Number Line

Distribute an index card to each student that is labeled with an integer ranging from -10 to 10. Create enough cards based on the class size. Have students stand or place their index cards on the number line one at a time. When placing their numbers, students should start at zero and move in the direction of their numbers, counting out loud. Pose discussion questions after the exercise.

Scaffolding:
- Remind students how to locate a negative number on the number line from Lesson 1.
- For advanced learners, provide positive and negative fractions. Pose the last three questions for inquiry only.

Discuss the following:

- What patterns do you see with the numbers on the number line?
  - *For each number to the right of zero, there is a corresponding number the same distance from zero to the left.*
- What does zero represent on the number line?
  - *Zero represents the reference point when locating a point on the number line. It also represents the separation of positive numbers from negative numbers.*
- What is the relationship between any two opposite numbers and zero on the number line?
  - *Opposite numbers are the same distance from zero, but they are on opposite sides of zero.*

Read and display the statement below to the class, and then model Example 1.

> **Exercise 1: Walk the Number Line**
>
> 1. Each nonzero integer has an opposite, denoted $-a$; $-a$ and a are opposites if they are on opposite sides of zero and the same distance from zero on the number line.

Example 1 (5 minutes): Every Number Has an Opposite

Read the example out loud. Model how to graph both points, and explain how they are related.

> **Example 1: Every Number Has an Opposite**
>
> Locate the number 8 and its opposite on the number line. Explain how they are related to zero.
>
>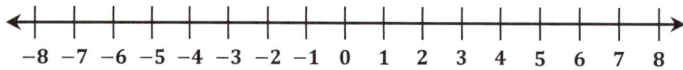

- First, start at zero, and move 8 units to the right to locate positive 8. So, the opposite of 8 must be 8 units to the left of zero. What number is 8 units to the left of zero?
  - -8
- 8 and -8 are the same distance from zero. Since both numbers are the same distance from zero but on opposite sides of zero on the number line, they are opposites.

A STORY OF RATIOS — Lesson 4 — 6•3

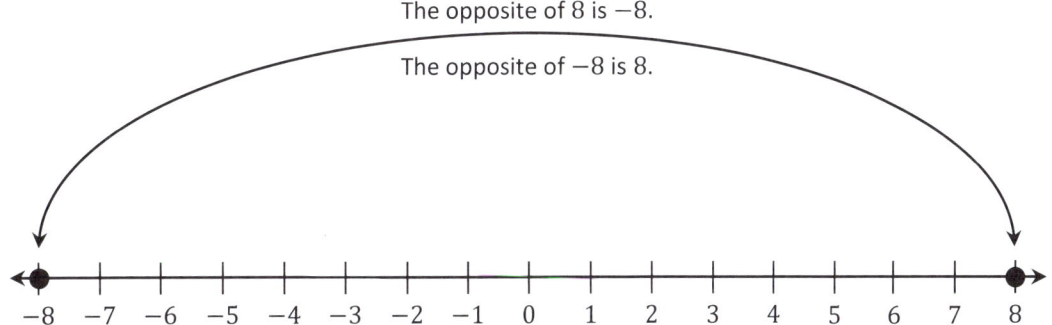

Exercises 2–3 (5 minutes)

Students work independently to answer the following questions. Allow 2–3 minutes for review as a whole group.

Exercises 2–3

2. Locate and label the opposites of the numbers on the number line.
   a. 9
   b. −2
   c. 4
   d. −7

   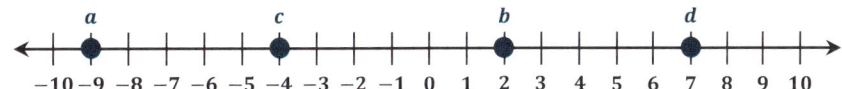

3. Write the integer that represents the opposite of each situation. Explain what zero means in each situation.

   a. 100 feet above sea level

   −100; zero represents sea level.

   b. 32°C below zero

   32; zero represents 0 degrees Celsius.

   c. A withdrawal of $25

   25; zero represents no change, where no withdrawal or deposit is made.

Lesson 4: The Opposite of a Number

A STORY OF RATIOS • Lesson 4 • 6•3

Example 2 (8 minutes): A Real-World Example

The purpose of this example is to show students how to graph opposite integers given a real-world situation. In pairs, have students read the problem aloud to each other. Instruct students to circle any words that might be important to solve the problem. Pose questions to the class while guiding students as a whole group through the example.

Example 2: A Real-World Example

Maria decides to take a walk along Central Avenue to purchase a book at the bookstore. On her way, she passes the Furry Friends Pet Shop and goes in to look for a new leash for her dog. Furry Friends Pet Shop is seven blocks west of the bookstore. She leaves Furry Friends Pet Shop and walks toward the bookstore to look at some books. After she leaves the bookstore, she heads east for seven blocks and stops at Ray's Pet Shop to see if she can find a new leash at a better price. Which location, if any, is the farthest from Maria while she is at the bookstore?

Determine an appropriate scale, and model the situation on the number line below.

Answers will vary.

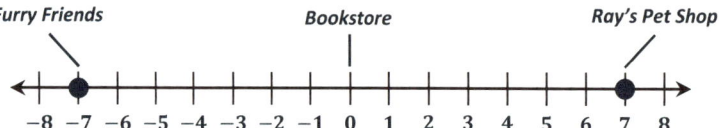

Explain your answer. What does zero represent in the situation?

The pet stores are the same distance from Maria, who is at the bookstore. They are each 7 blocks away but in opposite directions. In this example, zero represents the bookstore.

Discuss the following:

- How did you determine an appropriate scale for the situation if all blocks in the city are the same length?
    - *Because both stores are seven blocks in opposite directions, I knew that I could count by ones since the numbers are not that large.*
- Where would the bookstore be located on the number line?
    - *The bookstore would be located at zero.*
- Where would Ray's Pet Shop be located on the number line? Explain.
    - *It would be seven units to the right of zero because it is seven blocks east of the bookstore.*
- What integer represents this situation?
    - *7*
- Where would Furry Friends be located on the number line? Explain.
    - *It would be seven units to the left of zero because it is seven blocks west of the bookstore.*
- What integer represents this situation?
    - *−7*
- What do you notice about the distance between both stores from the bookstore?
    - *Both stores are the same distance from the bookstore but in opposite directions.*

MP.6

Students should practice clarifying any misconceptions about how to represent these situations as integers.

- "Seven blocks to the left" would not be written as "−7 blocks from the bookstore" or "−7 units from 0."
- Positive numbers are counting numbers and do not have a sign.

Exercises 4–6 (5 minutes)

Students work independently to answer the following questions. Allow 2–3 minutes for review as a whole group.

> **Exercises 4–6**
>
> Read each situation carefully, and answer the questions.
>
> 4. On a number line, locate and label a credit of $15 and a debit for the same amount from a bank account. What does zero represent in this situation?
>
> *Zero represents no change in the balance.*
>
>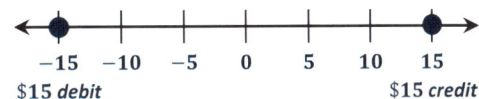
>
> 5. On a number line, locate and label 20°C below zero and 20°C above zero. What does zero represent in this situation?
>
> *Zero represents 0°C.*
>
>
>
> 6. A proton represents a positive charge. Write an integer to represent 5 protons. An electron represents a negative charge. Write an integer to represent 3 electrons.
>
> *5 protons: 5*
>
> *3 electrons: −3*

Closing (2 minutes)

- What is the relationship between any number and its opposite when plotted on a number line?
  - *A nonzero number and its opposite are both the same distance away from zero on a number line, but they are on opposite sides of zero.*
- How would you use this relationship to locate the opposite of a given number on the number line?
  - *I would use the given number to find the distance from zero on the opposite side.*
- Will this process work when finding the opposite of zero?
  - *No, because zero is its own opposite.*

Exit Ticket (5 minutes)

Lesson 4: The Opposite of a Number

Lesson 4: The Opposite of a Number

Exit Ticket

In a recent survey, a magazine reported that the preferred room temperature in the summer is 68°F. A wall thermostat, like the ones shown below, tells a room's temperature in degrees Fahrenheit.

Sarah's Upstairs Bedroom   Downstairs Bedroom

a. Which bedroom is warmer than the recommended room temperature?

b. Which bedroom is cooler than the recommended room temperature?

c. Sarah notices that her room's temperature is 4°F above the recommended temperature, and the downstairs bedroom's temperature is 4°F below the recommended temperature. She graphs 72 and 64 on a vertical number line and determines they are opposites. Is Sarah correct? Explain.

d. After determining the relationship between the temperatures, Sarah now decides to represent 72°F as 4 and 64°F as −4 and graphs them on a vertical number line. Graph 4 and −4 on the vertical number line on the right. Explain what zero represents in this situation.

A STORY OF RATIOS — Lesson 4 — 6•3

Exit Ticket Sample Solutions

In a recent survey, a magazine reported that the preferred room temperature in the summer is 68°F. A wall thermostat, like the ones shown below, tells a room's temperature in degrees Fahrenheit.

Sarah's Upstairs Bedroom   Downstairs Bedroom

a. Which bedroom is warmer than the recommended room temperature?

The upstairs bedroom is warmer than the recommended room temperature.

b. Which bedroom is cooler than the recommended room temperature?

The downstairs bedroom is cooler than the recommended room temperature.

c. Sarah notices that her room's temperature is 4°F above the recommended temperature, and the downstairs bedroom's temperature is 4°F below the recommended temperature. She graphs 72 and 64 on a vertical number line and determines they are opposites. Is Sarah correct? Explain.

No. Both temperatures are positive numbers and not the same distance from 0, so they cannot be opposites. Both numbers have to be the same distance from zero, but one has to be above zero, and the other has to be below zero in order to be opposites.

d. After determining the relationship between the temperatures, Sarah now decides to represent 72°F as 4 and 64°F as −4 and graphs them on a vertical number line. Graph 4 and −4 on the vertical number line on the right. Explain what zero represents in this situation.

Zero represents the recommended room temperature of 68°F. Zero could also represent not being above or below the recommended temperature.

Problem Set Sample Solutions

1. Find the opposite of each number, and describe its location on the number line.

   a. −5

   The opposite of −5 is 5, which is 5 units to the right of (or above) 0.

   b. 10

   The opposite of 10 is −10, which is 10 units to the left of (or below) 0.

   c. −3

   The opposite of −3 is 3, which is 3 units to the right of (or above) 0.

Lesson 4: The Opposite of a Number

d. 15

   The opposite of 15 is −15, which is 15 units to the left of (or below) 0.

2. Write the opposite of each number, and label the points on the number line.

   a. Point A: the opposite of 9

      −9

   b. Point B: the opposite of −4

      4

   c. Point C: the opposite of −7

      7

   d. Point D: the opposite of 0

      0

   e. Point E: the opposite of 2

      −2

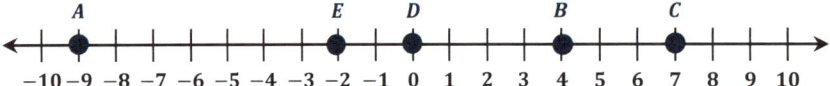

3. Study the first example. Write the integer that represents the opposite of each real-world situation. In words, write the meaning of the opposite.

   a. An atom's positive charge of 7   —7, *an atom's negative charge of 7*

   b. A deposit of $25   —25, *a withdrawal of* $25

   c. 3,500 feet below sea level   *3,500, 3,500 feet above sea level*

   d. A rise of 45°C   —45, *a decrease of* 45°C

   e. A loss of 13 pounds   *13, a gain of 13 pounds*

4. On a number line, locate and label a credit of $38 and a debit for the same amount from a bank account. What does zero represent in this situation?

   Zero represents no change in the balance.

   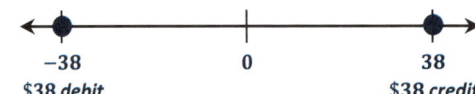

5. On a number line, locate and label 40°C below zero and 40°C above zero. What does zero represent in this situation?

   Zero represents 0°C.

   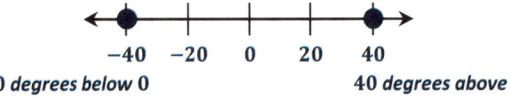

A STORY OF RATIOS

Lesson 5 6•3

Lesson 5: The Opposite of a Number's Opposite

Student Outcomes

- Students understand that, for instance, the opposite of -5 is denoted $-(-5)$ and is equal to 5. In general, they know that the opposite of the opposite is the original number; for example, $-(-a) = a$.
- Students locate and position opposite numbers on a number line.

Classwork

Opening Exercise (7 minutes)

Students work independently for 5 minutes to complete the following review problems and then review as a class.

Opening Exercise

a. Locate the number -2 and its opposite on the number line below.

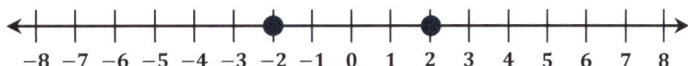

b. Write an integer that represents each of the following.

   i. 90 feet below sea level

   -90

   ii. $100 of debt

   -100

   iii. 2°C above zero

   2

c. Joe is at the ice cream shop, and his house is 10 blocks north of the shop. The park is 10 blocks south of the ice cream shop. When he is at the ice cream shop, is Joe closer to the park or his house? How could the number zero be used in this situation? Explain.

   He is the same distance from his house and the park because both are located 10 blocks away from the ice cream shop but in opposite directions. In this situation, zero represents the location of the ice cream shop.

Lesson 5: The Opposite of a Number's Opposite

Lesson 5

Example 1 (8 minutes): The Opposite of an Opposite of a Number

- What is the opposite of the opposite of 8? How can we illustrate this number on a number line? *(Example student responses are listed below.)*

Before starting the example, allow students to discuss their predictions in groups for one minute. Choose a few students to share their responses with the rest of the class.

> **Example 1: The Opposite of an Opposite of a Number**
>
> What is the opposite of the opposite of 8? How can we illustrate this number on a number line?
>
> a. What number is 8 units to the right of 0? __8__
>
> b. How can you illustrate locating the opposite of 8 on this number line?
>
> *We can illustrate the opposite of 8 on the number line by counting 8 units to the left of zero rather than to the right of zero.*
>
> c. What is the opposite of 8? __−8__
>
> d. Use the same process to locate the opposite of −8. What is the opposite of −8? __8__
>
>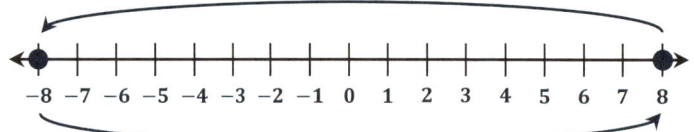
>
> e. The opposite of an opposite of a number is __the original number__.

Scaffolding:

For visual learners, use the curved arrows initially to develop conceptual understanding of how to find the opposite of an opposite of a number.

Example 2 (8 minutes): Writing the Opposite of an Opposite of a Number

In this example, students use the "−" symbol to indicate "the opposite of a number." Display the task, "Explain why $-(-5) = 5$" on the board. A "−" symbol means "the opposite of a number."

$$-(-5) = 5$$

Since the opposite of 5 is negative 5 and the opposite of negative 5 is positive 5, then $-(-5) = 5$.

Pose the following questions to the class to check for understanding.

- What is the opposite of −6?
  - 6
- What is the opposite of the opposite of 10?
  - 10
- How would you write the opposite of the opposite of −12?
  - The opposite of −12: $-(-12) = 12$. The opposite of 12: $-(12) = -12$. So, $-(-(-12)) = -12$.
- What does a "−" symbol mean?
  - It can mean the opposite of a number or indicate a negative number.

Lesson 5: The Opposite of a Number's Opposite

A STORY OF RATIOS　　　　　　　　　　　　　　　　　　　　　　　　　　　　　Lesson 5　6•3

- What is the opposite of the opposite of a debit of $12?
  □ A debit of $12 is -12. The opposite of -12 is 12, or a credit of $12. The opposite of a credit of $12 is a debit of $12.
- In general, the opposite of the opposite of a number is the original number; for example, $-(-a) = a$.

Exercises 1–3 (12 minutes)

Students work in groups of 3–4 to find the opposite of an opposite of a set of given numbers. Allow 4–5 minutes for review as a whole class. Distribute a card with an integer on it to each group. (Use sticky notes or index cards.) Students complete the table using all the cards in their group.

Exercises

Complete the table using the cards in your group.

Person	Card (a)	Opposite of Card ($-a$)	Opposite of Opposite of Card $-(-a)$
Jackson	4	$-(4) = -4$	$-(-4) = 4$
DeVonte	150	$-(150) = -150$	$-(-150) = 150$
Cheryl	-6	$-(-6) = 6$	$-(-(-6)) = -6$
Toby	-9	$-(-9) = 9$	$-(-(-9)) = -9$

1. Write the opposite of the opposite of -10 as an equation.

   The opposite of -10: $-(-10) = 10$; the opposite of 10: $-(10) = -10$. Therefore, $\left(-(-(-10))\right) = -10$.

2. In general, the opposite of the opposite of a number is the **original number**.

3. Provide a real-world example of this rule. Show your work.

   Answers will vary. The opposite of the opposite of 100 feet below sea level is 100 feet below sea level.
   -100 is 100 feet below sea level.
   $-(-100) = 100$, the opposite of -100
   $-(100) = -100$, the opposite of 100

Closing (3 minutes)

- What is the opposite of an opposite of a number? Support your answer with an example.
  □ The opposite of an opposite of a number is the original number. The opposite of the opposite of -6 is -6 because the opposite of -6 is 6. The opposite of 6 is -6.
- What is the relationship between the location of a nonzero number on the number line and the location of its opposite on the number line?
  □ A number and its opposite are located the same distance from 0 on a number line but on opposite sides of 0.

Exit Ticket (7 minutes)

Lesson 5: 　 The Opposite of a Number's Opposite 　 47

A STORY OF RATIOS   Lesson 5   6•3

Name _____   Date _____

Lesson 5: The Opposite of a Number's Opposite

Exit Ticket

1. Jane completes several example problems that ask her to the find the opposite of the opposite of a number, and for each example, the result is a positive number. Jane concludes that when she takes the opposite of the opposite of any number, the result will always be positive. Is Jane correct? Why or why not?

2. To support your answer from the previous question, create an example, written as an equation. Illustrate your example on the number line below.

Exit Ticket Sample Solutions

1. Jane completes several example problems that ask her to the find the opposite of the opposite of a number, and for each example, the result is a positive number. Jane concludes that when she takes the opposite of the opposite of any number, the result will always be positive. Is Jane correct? Why or why not?

   She is not correct. The opposite of the opposite of a number is the original number. So, if Jane starts with a negative number, she will end with a negative number.

2. To support your answer from the previous question, create an example, written as an equation. Illustrate your example on the number line below.

   If Jane starts with -7, the opposite of the opposite of -7 is written as $-(-(-7)) = -7$ or the opposite of -7: $-(-7) = 7$; the opposite of 7: $-(7) = -7$.

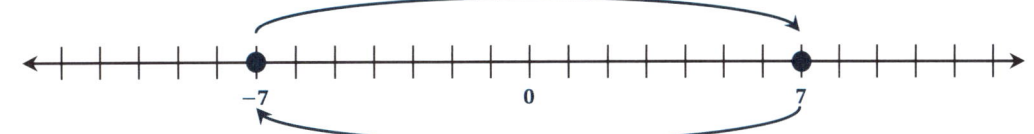

Problem Set Sample Solutions

1. Read each description carefully, and write an equation that represents the description.

   a. The opposite of negative seven

      $-(-7) = 7$

   b. The opposite of the opposite of twenty-five

      $-(-(25)) = 25$

   c. The opposite of fifteen

      $-(15) = -15$

   d. The opposite of negative thirty-six

      $-(-36) = 36$

2. Jose graphed the opposite of the opposite of 3 on the number line. First, he graphed point P on the number line 3 units to the right of zero. Next, he graphed the opposite of P on the number line 3 units to the left of zero and labeled it K. Finally, he graphed the opposite of K and labeled it Q.

   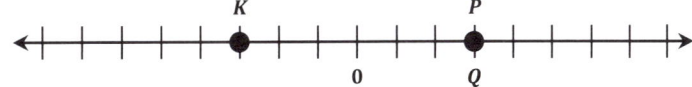

   a. Is his diagram correct? Explain. If the diagram is not correct, explain his error, and correctly locate and label point Q.

   Yes, his diagram is correct. It shows that point P is 3 because it is 3 units to the right of zero. The opposite of 3 is -3, which is point K (3 units to the left of zero). The opposite of -3 is 3, so point Q is 3 units to the right of zero.

Lesson 5: The Opposite of a Number's Opposite

b. Write the relationship between the points:

P and K _They are opposites._

K and Q _They are opposites._

P and Q _They are the same._

3. Read each real-world description. Write the integer that represents the opposite of the opposite. Show your work to support your answer.

   a. A temperature rise of 15 degrees Fahrenheit

   -15 _is the opposite of_ 15 _(fall in temperature)._
   15 _is the opposite of_ -15 _(rise in temperature)._
   $-(-(15)) = 15$

   b. A gain of 55 yards

   -55 _is the opposite of_ 55 _(loss of yards)._
   55 _is the opposite of_ -55 _(gain of yards)._
   $-(-(55)) = 55$

   c. A loss of 10 pounds

   10 _is the opposite of_ -10 _(gain of pounds)._
   -10 _is the opposite of_ 10 _(loss of pounds)._
   $-(-(-10)) = -10$

   d. A withdrawal of $\$2,000$

   $2,000$ _is the opposite of_ $-2,000$ _(deposit)._
   $-2,000$ _is the opposite of_ $2,000$ _(withdrawal)._
   $-(-(-2,000)) = -2,000$

4. Write the integer that represents the statement. Locate and label each point on the number line below.

   a. The opposite of a gain of 6 -6
   b. The opposite of a deposit of $\$10$ -10
   c. The opposite of the opposite of 0 0
   d. The opposite of the opposite of 4 4
   e. The opposite of the opposite of a loss of 5 -5

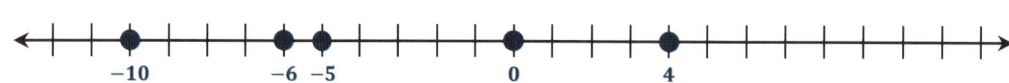

Lesson 5: The Opposite of a Number's Opposite

A STORY OF RATIOS   Lesson 6   6•3

 Lesson 6: Rational Numbers on the Number Line

Student Outcomes

- Students use number lines that extend in both directions and use 0 and 1 to locate integers and rational numbers on the number line. Students know that the sign of a nonzero rational number is positive or negative, depending on whether the number is greater than zero (positive) or less than zero (negative), and use an appropriate scale when graphing rational numbers on the number line.
- Students know that the opposites of rational numbers are similar to the opposites of integers. Students know that two rational numbers have opposite signs if they are on different sides of zero and that they have the same sign if they are on the same side of zero on the number line.

Classwork

Opening Exercise (5 minutes)

Students work independently for 5 minutes to review fractions and decimals.

Opening Exercise

a. Write the decimal equivalent of each fraction.

   i. $\frac{1}{2}$

      0.5

   ii. $\frac{4}{5}$

       0.8

   iii. $6\frac{7}{10}$

        6.70

b. Write the fraction equivalent of each decimal.

   i. 0.42

      $\frac{42}{100} = \frac{21}{50}$

   ii. 3.75

       $3\frac{75}{100} = 3\frac{3}{4}$

   iii. 36.90

        $36\frac{90}{100} = 36\frac{9}{10}$

Scaffolding:
- Use polling software to elicit immediate feedback from the class to engage all learners.
- Display each problem one at a time, and use personal white boards for kinesthetic learners.

Scaffolding:
- Use edges of square tiles on the floor as a number line to illustrate how to connect segments of equal length for visual and kinesthetic learners.
- Provide green and red pencils to help with modeling the example for visual learners.

Lesson 6:   Rational Numbers on the Number Line   51

This work is derived from Eureka Math ™ and licensed by Great Minds. ©2015 Great Minds. eureka-math.org
G6-M3-TE-B3-1.3.0-07.2015

A STORY OF RATIOS — Lesson 6 — 6•3

Example 1 (10 minutes): Graphing Rational Numbers

The purpose of this example is to show students how to graph non-integer rational numbers on a real number line. Students complete the example by following along with the teacher.

- Locate and graph the number $\frac{3}{10}$ and its opposite on a number line.

Before modeling the example, review graphing a fraction on the number line to the whole class by first reviewing fraction definitions with respect to the number line.

> **Example 1: Graphing Rational Numbers**
>
> If b is a nonzero whole number, then the unit fraction $\frac{1}{b}$ is located on the number line by dividing the segment between 0 and 1 into b segments of equal length. One of the b segments has 0 as its left end point; the right end point of this segment corresponds to the unit fraction $\frac{1}{b}$.

- Since the number is a **rational number**, a number that can be represented as a fraction, determine how the number line should be scaled.[1]
- First, divide the number line into two halves to represent positive and negative numbers.

Have students complete this task on their student pages.

- Next, divide the right half of the number line segment between 0 and 1 into ten segments of equal length; each segment has a length of $\frac{1}{10}$.

Students divide their number lines into ten equal segments as shown. Check for accuracy.

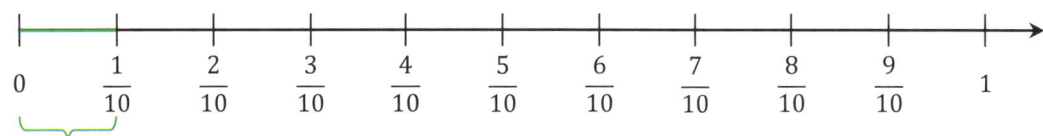

- There are 10 equal segments. Each segment has a length of $\frac{1}{10}$. The first segment has 0 as its left end point, and the right end point corresponds to $\frac{1}{10}$.

---

[1]Supplemental Exercise:
- Have four students each stand in a square floor tile forming a straight line facing the class. Give each student a number to tie around his neck: $0, \frac{1}{10}, \frac{2}{10}$, or $\frac{3}{10}$. (Use index cards or construction paper.)
- Ask a fifth student to assist by giving one end of a ball of string to the person at 0. This person holds one end of the string and passes the rest to the person to the left. (So the class sees it moving to the right.)
- As the string gets passed down the line, each person announces her number, " $\frac{1}{10}, \frac{2}{10}, \frac{3}{10}$ " stopping at $\frac{3}{10}$.
- The assistant cuts the string at $\frac{3}{10}$ and gives that end of the string to the person holding $\frac{3}{10}$, making one segment of length $\frac{3}{10}$.
- Have students turn over their numbers to reveal their opposites and rearrange themselves to represent the opposite of $\frac{3}{10}$ using the same process. This time, students pass the string to the right. (So the class sees it moving to the left.)

> The fraction $\frac{a}{b}$ is located on the number line by joining a segments of length $\frac{1}{b}$ so that (1) the left end point of the first segment is 0, and (2) the right end point of each segment is the left end point of the next segment. The right end point of the last segment corresponds to the fraction $\frac{a}{b}$.

- To locate the number $\frac{a}{b}$ on a number line, students should divide the interval between zero and 1 into b equal parts. Starting at 0, move along the number line a number of times.

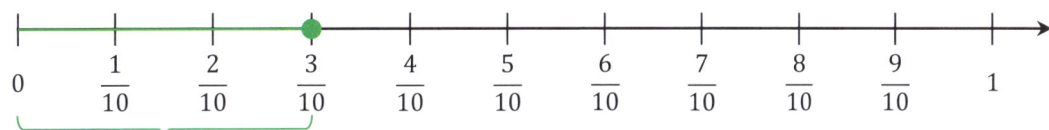

- There are ten equal segments. Each segment has a length of $\frac{1}{10}$. The first segment has a 0 as its left end point, and the right end point of the third segment corresponds to $\frac{3}{10}$. The point is located at $\frac{3}{10}$.

- The opposite of $\frac{3}{10}$ is located the same distance from zero as $\frac{3}{10}$ but in the opposite direction or to the left. Using your knowledge of opposites, what rational number represents the opposite of $\frac{3}{10}$?
  - $-\frac{3}{10}$

- To locate the opposite of $\frac{3}{10}$ on the number line, divide the interval between zero and -1 into ten equal segments. Starting at zero, how far would we move to locate the opposite of $\frac{3}{10}$, and in what direction?
  - *We would move 3 units to the left of zero because that is the same distance but opposite direction we moved to plot the point $\frac{3}{10}$.*

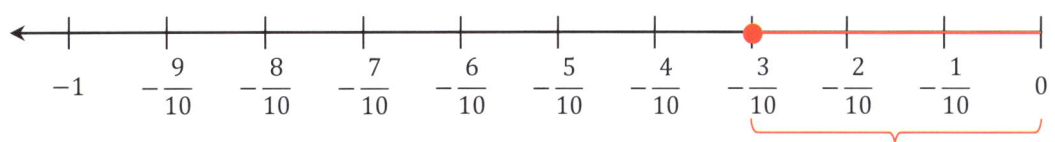

- There are ten equal segments. Each segment has a length of $\frac{1}{10}$. Three consecutive segments, starting at 0 and moving to the left, would have a total length of $\frac{3}{10}$. The point is located at $-\frac{3}{10}$.

- Counting three consecutive segments of length of $\frac{1}{10}$ from 0 moving to the left and taking the end point of the last segment corresponds to the number $-\frac{3}{10}$. Therefore, the opposite of $\frac{3}{10}$ is $-\frac{3}{10}$.

> Locate and graph the number $\frac{3}{10}$ and its opposite on a number line.
>
>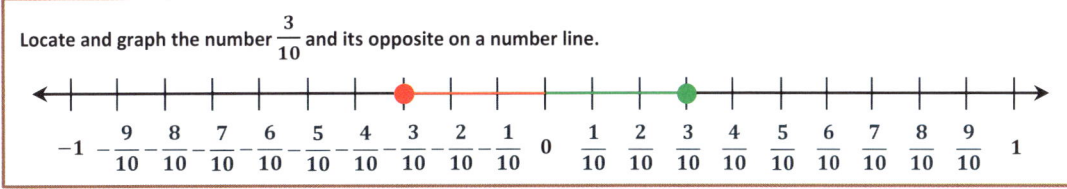

Lesson 6: Rational Numbers on the Number Line

A STORY OF RATIOS   Lesson 6   6•3

Exercise 1 (5 minutes)

Students work independently to practice graphing a non-integer rational number and its opposite on the number line. Allow 2–3 minutes for review as a whole group.

> **Exercise 1**
>
> Use what you know about the point $-\frac{7}{4}$ and its opposite to graph both points on the number line below. The fraction $-\frac{7}{4}$ is located between which two consecutive integers? Explain your reasoning.
>
>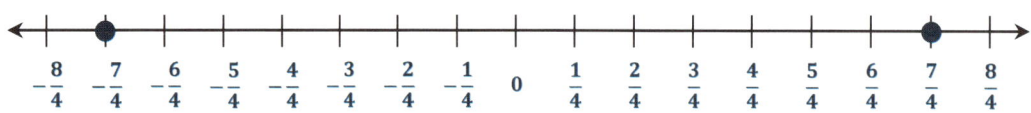
>
> On the number line, each segment will have an equal length of $\underline{\frac{1}{4}}$. The fraction is located between $\underline{-1}$ and $\underline{-2}$.
>
> **Explanation:**
>
> $\frac{7}{4}$ is the opposite of $-\frac{7}{4}$. It is the same distance from zero but on the opposite side of zero. Since $-\frac{7}{4}$ is to the left of zero, $\frac{7}{4}$ is to the right of zero. The original fraction is located between -2 (or $-\frac{8}{4}$) and -1 (or $-\frac{4}{4}$).

Example 2 (7 minutes): Rational Numbers and the Real World

Display the following vertical number line model on the board. Students are to follow along in their student materials to answer the questions. Pose additional questions to the class throughout the example.

> **Example 2: Rational Numbers and the Real World**
>
> The water level of a lake rose 1.25 feet after it rained. Answer the following questions using the number line below.
>
> a. Write a rational number to represent the situation.
>
> 1.25 or $1\frac{1}{4}$
>
> b. What two integers is 1.25 between on a number line?
>
> **1 and 2**
>
> c. Write the length of each segment on the number line as a decimal and a fraction.
>
> 0.25 and $\frac{1}{4}$
>
> d. What will be the water level after it rained? Graph the point on the number line.
>
> 1.25 *feet above the original lake level*
>
>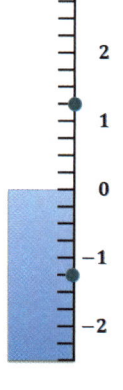

Lesson 6: Rational Numbers on the Number Line

> e. After two weeks have passed, the water level of the lake is now the opposite of the water level when it rained. What will be the new water level? Graph the point on the number line. Explain how you determined your answer.
>
> *The water level would be 1.25 feet below the original lake level. If the water level was 1.25, the opposite of 1.25 is -1.25.*
>
> f. State a rational number that is not an integer whose value is less than 1.25, and describe its location between two consecutive integers on the number line.
>
> *Answers will vary. A rational number whose value is less than 1.25 is 0.75. It would be located between 0 and 1 on a number line.*

Possible discussion questions:

- What units are we using to measure the water level?
  - Feet
- What was the water level after the rain? How do you know?
  - If zero represents the original water level on the number line, the water level after rain is 1.25 feet. From 0 to 1, there are four equal segments. This tells me that the scale is $\frac{1}{4}$. The top of the water is represented on the number line at one mark above 1, which represents $\frac{5}{4}$ feet or 1.25 feet.
- What strategy could we use to determine the location of the water level on the number line after it rained?
  - I started at 0 and counted by $\frac{1}{4}$ for each move. I counted $\frac{1}{4}$ five times to get $\frac{5}{4}$, which is equivalent to $1\frac{1}{4}$ and 1.25. I know the number is positive because I moved up. Since the measurements are in feet, the answer is 1.25 feet.
- For the fraction $\frac{5}{4}$, what is the value of the numerator and denominator?
  - The numerator is 5, and the denominator is 4.
- What do the negative numbers represent on the number line?
  - They represent the number of feet below the original lake level.

Exercise 2 (10 minutes)

Students are seated in groups of three or four. Distribute one sheet of grid paper and a ruler to each student. Each group completes the following tasks:

1. Write a real-world story problem using a rational number and its opposite.
2. Create a horizontal or vertical number line diagram to represent your situation.
   a. Determine an appropriate scale, and label the number line.
   b. Write the units of measurement (if needed).
   c. Graph the rational number and its opposite that represent the situation.
3. Describe what points 0 and the opposite number represent on the number line.
4. Name a rational number to the left and right of the rational number you initially chose.

> *Scaffolding:*
> - *Project the directions for the example as a way for groups to make sure they are completing all task requirements.*
> - *Have students write their story problems and draw their number lines on large wall grid paper.*
> - *Hang posters around the room to use as a gallery walk for students who finish their Exit Tickets early, or use them as review for the Mid-Module Assessment later in the module.*

Lesson 6: Rational Numbers on the Number Line

> **Exercise 2**
>
> **Our Story Problem**
>
> *Answers will vary.*
>
> Melissa and Samantha weigh the same amount. Melissa gained 5.5 pounds last month, while Samantha lost the same amount Melissa gained.
>
>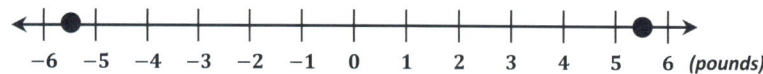
>
> - Our Scale: 1
> - Our Units: Pounds
> - Description: On the number line, zero represents Melissa and Samantha's original weight. The point −5.5 represents the change in Samantha's weight. The amount lost is 5.5 pounds.
> - Other Information: A rational number to the left of 5.5 is 4.5. A rational number to the right of 5.5 is 5.75.

Closing (2 minutes)

- How is graphing the number $\frac{4}{3}$ on a number line similar to graphing the number 4 on a number line?
  - *When graphing each number, you start at zero and move to the right (or up) 4 units.*
- How is graphing the number $\frac{4}{3}$ on a number line different from graphing the number 4 on a number line?
  - *When we graph 4, the unit length is one, and when we graph $\frac{4}{3}$, the unit length is $\frac{1}{3}$.*
- On a vertical number line, describe the location of the rational number that represents −5.1 and its opposite.
  - *The number −5.1 would be 5.1 units below zero because it is negative. Its opposite, 5.1, would be 5.1 units above zero because it is positive.*

Exit Ticket (6 minutes)

Name _____   Date _____

Lesson 6: Rational Numbers on the Number Line

Exit Ticket

Use the number line diagram below to answer the following questions.

1. What is the length of each segment on the number line?

2. What number does point K represent?

3. What is the opposite of point K?

4. Locate the opposite of point K on the number line, and label it point L.

5. In the diagram above, zero represents the location of Martin Luther King Middle School. Point K represents the library, which is located to the east of the middle school. In words, create a real-world situation that could represent point L, and describe its location in relation to 0 and point K.

A STORY OF RATIOS   Lesson 6 6•3

Exit Ticket Sample Solutions

Use the number line diagram below to answer the following questions.

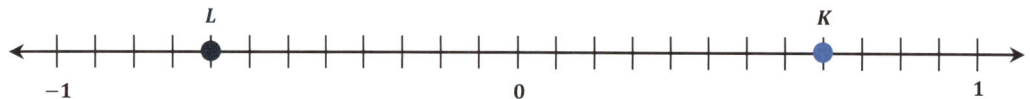

1. What is the length of each segment on the number line?

   $\dfrac{1}{12}$

2. What number does point K represent?

   $\dfrac{8}{12}$, or $\dfrac{2}{3}$

3. What is the opposite of point K?

   $-\dfrac{8}{12}$, or $-\dfrac{2}{3}$

4. Locate the opposite of point K on the number line, and label it point L.

5. In the diagram above, zero represents the location of Martin Luther King Middle School. Point K represents the library, which is located to the east of the middle school. In words, create a real-world situation that could represent point L, and describe its location in relation to 0 and point K.

   Answers may vary. Point L is $\dfrac{8}{12}$ units to the left of 0, so it is a negative number. Point L represents the recreation center, which is located $\dfrac{8}{12}$ mile west of Martin Luther King Middle School. This means that the recreation center and library are the same distance from the middle school but in opposite directions because the opposite of $\dfrac{8}{12}$ is $-\dfrac{8}{12}$.

Problem Set Sample Solutions

Students gain additional practice with graphing rational numbers on the number line.

1. In the space provided, write the opposite of each number.

   a. $\dfrac{10}{7}$ $-\dfrac{10}{7}$

   b. $-\dfrac{5}{3}$ $\dfrac{5}{3}$

   c. 3.82 -3.82

   d. $-6\dfrac{1}{2}$ $6\dfrac{1}{2}$

58   Lesson 6:   Rational Numbers on the Number Line

2. Choose a non-integer between 0 and 1. Label it point A and its opposite point B on the number line. Write values below the points.

   (Answers may vary.)

   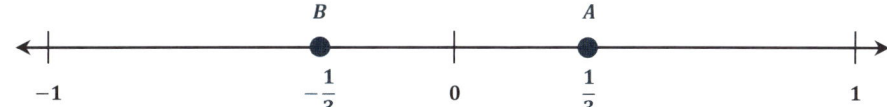

   a. To draw a scale that would include both points, what could be the length of each segment?

      Answers may vary. $\frac{1}{3}$

   b. In words, create a real-world situation that could represent the number line diagram.

      Answers may vary. Starting at home, I ran $\frac{1}{3}$ mile. My brother ran $\frac{1}{3}$ mile from home in the opposite direction.

3. Choose a value for point P that is between -6 and -7.

   Answers may vary. $-\frac{13}{2}, -6.25, -6.8$

   a. What is the opposite of point P?

      Answers may vary. $\frac{13}{2}, 6.25, 6.8$

   b. Use the value from part (a), and describe its location on the number line in relation to zero.

      $\frac{13}{2}$ is the same distance as $-\frac{13}{2}$ from zero but to the right. $\frac{13}{2}$ is $6\frac{1}{2}$ units to the right of (or above) zero.

   c. Find the opposite of the opposite of point P. Show your work, and explain your reasoning.

      The opposite of an opposite of the number is the original number. If point P is located at $-\frac{13}{2}$, then the opposite of the opposite of point P is located at $-\frac{13}{2}$. The opposite of $-\frac{13}{2}$ is $\frac{13}{2}$. The opposite of $\frac{13}{2}$ is $-\frac{13}{2}$.
      $-\left(-\frac{13}{2}\right) = \frac{13}{2}$ *and* $-\left(-\left(-\frac{13}{2}\right)\right) = -\frac{13}{2}$

4. Locate and label each point on the number line. Use the diagram to answer the questions.

Jill lives one block north of the pizza shop.

Janette's house is $\frac{1}{3}$ block past Jill's house.

Jeffrey and Olivia are in the park $\frac{4}{3}$ blocks south of the pizza shop.

Jenny's Jazzy Jewelry Shop is located halfway between the pizza shop and the park.

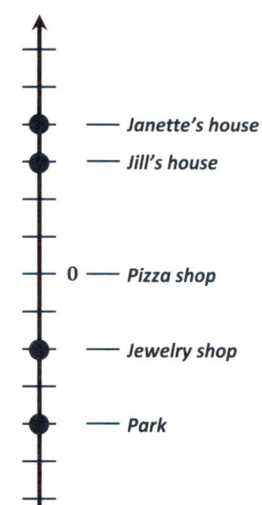

a. Describe an appropriate scale to show all the points in this situation.

   An appropriate scale would be $\frac{1}{3}$ because the numbers given in the example all have denominators of 3. I would divide the number line into equal segments of $\frac{1}{3}$.

b. What number represents the location of Jenny's Jazzy Jewelry Shop? Explain your reasoning.

   The number is $-\frac{2}{3}$. I got my answer by finding the park first. It is 4 units below 0. Since the jewelry shop is halfway between the pizza shop and the park, half of 4 is 2. Then, I moved 2 units down on the number line since the shop is south of the pizza shop before the park.

A STORY OF RATIOS

6 GRADE

Mathematics Curriculum

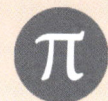

GRADE 6 • MODULE 3

Topic B
Order and Absolute Value

6.NS.C.6c, 6.NS.C.7

Focus Standards:	6.NS.C.6c	Understand a rational number as a point on the number line. Extend number line diagrams and coordinate axes familiar from previous grades to represent points on the line and in the plane with negative number coordinates.
		c. Find and position integers and other rational numbers on a horizontal or vertical number line diagram; find and position pairs of integers and other rational numbers on a coordinate plane.
	6.NS.C.7	Understand ordering and absolute value of rational numbers.
		a. Interpret statements of inequality as statements about the relative position of two numbers on a number line diagram. *For example, interpret $-3 > -7$ as a statement that -3 is located to the right of -7 on a number line oriented from left to right.*
		b. Write, interpret, and explain statements of order for rational numbers in real-world contexts. *For example, write $-3°C > -7°C$ to express the fact that $-3°C$ is warmer than $-7°C$.*
		c. Understand the absolute value of a rational number as its distance from 0 on the number line; interpret absolute value as magnitude for a positive or negative quantity in a real-world situation. *For example, for an account balance of -30 dollars, write $|-30| = 30$ to describe the size of the debt in dollars.*
		d. Distinguish comparisons of absolute value from statements about order. *For example, recognize that an account balance less than -30 dollars represents a debt greater than 30 dollars.*
Instructional Days:	7	
Lessons 7–8:	Ordering Integers and Other Rational Numbers (P, P)[1]	
Lesson 9:	Comparing Integers and Other Rational Numbers (P)	
Lesson 10:	Writing and Interpreting Inequality Statements Involving Rational Numbers (P)	
Lesson 11:	Absolute Value—Magnitude and Distance (P)	

---

[1]Lesson Structure Key: **P**-Problem Set Lesson, **M**-Modeling Cycle Lesson, **E**-Exploration Lesson, **S**-Socratic Lesson

Topic B: Order and Absolute Value

This work is derived from Eureka Math ™ and licensed by Great Minds. ©2015 Great Minds. eureka-math.org
G6-M3-TE-B3-1.3.0-07.2015

Topic B

Lesson 12: The Relationship Between Absolute Value and Order (P)

Lesson 13: Statements of Order in the Real World (P)

In Topic B, students focus on the ordering of rational numbers. They understand absolute value as a number's distance from zero on the number line (**6.NS.C.7**) and continue to find and position rational numbers on horizontal and vertical number lines (**6.NS.C.6c**). In Lessons 7–10, students compare and order integers and other rational numbers. They write, interpret, and explain statements of order for rational numbers in real-world contexts (**6.NS.C.7b**). For instance, if the temperature reached a low of $-8°F$ on Saturday evening and it is expected to be colder later Saturday night, students are able to arrive at a possible value for the temperature later that night, realizing that it would have to lie below or to the left of -8 on the number line. Students interpret inequality statements about the positioning of rational numbers with respect to one another (**6.NS.C.7a**) and recognize that if $a < b$, then $-a > -b$ because a number and its opposite are equal distances from zero.

In Lessons 11–13, students understand absolute value to be a number's distance from zero. They know opposite numbers are the same distance from zero and, therefore, have the same absolute value (**6.NS.C.7c**). Students interpret absolute value as magnitude and express their answers to real-world situations based on the context. For example, if a savings account statement shows a transaction amount of -400, students know that there must have been a 400 dollar withdrawal and that a withdrawal of more than 400 dollars is represented by a number less than -400 (**6.NS.C.7d**). Ordering rational numbers and the use of absolute value and the context of a situation culminates in Lesson 13. Students examine real-world scenarios and describe the relationship that exists among the rational numbers involved. They compare rational numbers and write statements of inequality based on the number line model using absolute value to determine, for instance, that a depth of 40 feet below sea level is 15 feet farther from sea level than a height of 25 feet above sea level.

Lesson 7: Ordering Integers and Other Rational Numbers

Student Outcomes

- Students write, interpret, and explain statements of order for rational numbers in the real world.
- Students recognize that if $a < b$, then $-a > -b$ because a number and its opposite are equal distances from zero. Students also recognize that moving along the horizontal number line to the right means the numbers are increasing.

Classwork

Opening Exercise (6 minutes): Guess My Integer and Guess My Rational Number

For this verbal warm-up exercise, students arrive at a rational number based on parameters provided by the teacher and their own questioning. Students are allowed to ask *no more than three questions* to help them figure out the number the teacher is thinking of. Students must raise their hands and wait to be called upon to ask a question. This is a whole-group activity, and once students know what the number is, they should raise their hands. After the activity is modeled several times, students in groups of two or three should try the activity among themselves, with one person thinking of a number and the others asking questions and guessing the number.

The game should increase in difficulty so that the first round involves an integer number, the second round involves a non-integer positive rational number, and the third round involves a non-integer negative rational number.

A sample dialogue for two rounds of the game is stated below.

Round 1: Guess My Integer

- I am thinking of an integer between -7 and 0. (The teacher is thinking of the number -1.)
    - *Student 1*: Is the number greater than -6?
- Yes
    - *Student 2*: Is the number less than -3?
- No
    - *Student 3*: Is the number -2?
- No
- You have asked me three questions. Raise your hand if you know the integer that I am thinking of.
    - -1
- Yes!

Round 2: Guess My Positive Non-Integer Rational Number

- I am thinking of a positive rational number between 1 and 2. (The teacher is thinking of the number $\frac{5}{4}$.)
    - *Student 1*: Is the number greater than $1\frac{1}{2}$?
- No
    - *Student 2*: Is the number less than 1.4?
- Yes
    - *Student 3*: Is the number less than 1.2?
- No
- You have asked me three questions. Raise your hand if you know the rational number that I am thinking of.
    - 1.3
- No
    - 1.35
- No
    - 1.25
- Yes!

Discussion (3 minutes)

- If we have two rational numbers that we want to graph on the same number line, how would we describe to someone where each number lies in comparison to the other number?
    - *The lesser number is to the left of (or below) the greater number. (Or the greater number is to the right of [or above] the lesser number.) Negative numbers are to the left of (or below) zero, and positive numbers are to the right of (or above) zero.*
- Describe the location on the number line of a number and its opposite.
    - *They are both the same distance from zero but on opposite sides of zero.*
- If we were to list the numbers from least to greatest, which would be farthest to the left on the number line? Why? Which would be farthest to the right? Why?
    - *The numbers on the number line decrease as you move to the left (or down) and increase as you move to the right (or up). So, the number that is the least would be farthest left (or down), and the number that is the greatest would be farthest right (or up).*

A STORY OF RATIOS

Lesson 7 6•3

Exercise 1 (5 minutes)

Allow time for a whole-group discussion to follow completion of the exercise.

> **Exercise 1**
>
> a. Graph the number 7 and its opposite on the number line. Graph the number 5 and its opposite on the number line.
>
>
>
> b. Where does 7 lie in relation to 5 on the number line?
>
> *On the number line, 7 is 2 units to the right of 5.*
>
> c. Where does the opposite of 7 lie on the number line in relation to the opposite of 5?
>
> *On the number line, -7 is 2 units to the left of -5.*
>
> d. I am thinking of two numbers. The first number lies to the right of the second number on a number line. What can you say about the location of their opposites? (If needed, refer to your number line diagram.)
>
> *On the number line, the opposite of the second number must lie to the right of the opposite of the first number. If we call the first number f and the second number s, then $-f$ and $-s$ will have the opposite order of f and s because $-f$ and $-s$ are opposites of f and s, so they lie on the opposite side of zero.*

Example 1 (4 minutes)

> *Scaffolding:*
>
> One potential strategy to use when tackling word problems is the RDW approach, which stands for **R**ead, **D**raw, and **W**rite.

 MP.2

> **Example 1**
>
> The record low temperatures for a town in Maine are $-20°F$ for January and $-19°F$ for February. Order the numbers from least to greatest. Explain how you arrived at the order.
>
> - *Read:* January: -20 and February: -19
> - *Draw:* Draw a number line model.
> - *Write:* Since -20 is farthest below zero and -19 is above -20 on the vertical number line, -20 is less than -19.
> - *Answer:* $-20, -19$
>
>

Lesson 7: Ordering Integers and Other Rational Numbers

A STORY OF RATIOS   Lesson 7   6•3

Exercises 2–4 (8 minutes)

Allow time to discuss these problems as a whole class.

> **Exercises 2–4**
>
> For each problem, order the rational numbers from least to greatest by first reading the problem, then drawing a number line diagram, and finally, explaining your answer.
>
> 2. Jon's time for running the mile in gym class is 9.2 minutes. Jacky's time is 9.18 minutes. Who ran the mile in less time?
>
> $9.18, 9.2$
>
>   *I drew a number line and graphed 9.2 and 9.18; 9.2 is to the right of 9.18. So, 9.18 is less than 9.2, which means Jacky ran the mile in less time than Jon.*
>
> 3. Mrs. Rodriguez is a teacher at Westbury Middle School. She gives bonus points on tests for outstanding written answers and deducts points for answers that are not written correctly. She uses rational numbers to represent the points. She wrote the following on students' papers: Student A: -2 points, Student B: -2.5 points. Did Student A or Student B perform worse on the test?
>
> $-2.5, -2$
>
>   *I drew a number line, and -2 and -2.5 are both to the left of zero, but -2.5 is to the left of -2. So, -2.5 is less than -2. That means Student B did worse than Student A.*
>
> 4. A carp is swimming approximately $8\frac{1}{4}$ feet beneath the water's surface, and a sunfish is swimming approximately $3\frac{1}{2}$ feet beneath the water's surface. Which fish is swimming farther beneath the water's surface?
>
> $-8\frac{1}{4}, -3\frac{1}{2}$
>
>   *I drew a vertical number line, and $-8\frac{1}{4}$ is farther below zero than $-3\frac{1}{2}$. So, $-8\frac{1}{4}$ is less than $-3\frac{1}{2}$, which means the carp is swimming farther beneath the water's surface.*

Example 2 (3 minutes)

> **Example 2**
>
> Henry, Janon, and Clark are playing a card game. The object of the game is to finish with the most points. The scores at the end of the game are Henry: -7, Janon: 0, and Clark: -5. Who won the game? Who came in last place? Use a number line model, and explain how you arrived at your answer.
>
> - *Read: Henry: -7, Janon: 0, and Clark: -5*
> - *Draw:*
>
>
>
> - *Explain: $-7, -5, 0$*
> - *Janon won the game, and Henry came in last place. I ordered the numbers on a number line, and -7 is farthest to the left. That means -7 is the smallest of the three numbers, so Henry came in last place. Next on the number line is -5, which is to the right of -7 but to the left of 0. Farthest to the right is 0; therefore, 0 is the greatest of the three numbers. This means Janon won the game.*

Exercises 5–6 (6 minutes)

> **Exercises 5–6**
>
> For each problem, order the rational numbers from least to greatest by first reading the problem, then drawing a number line diagram, and finally, explaining your answer.
>
> 5. Henry, Janon, and Clark are playing another round of the card game. Their scores this time are as follows: Clark: -1, Janon: -2, and Henry: -4. Who won? Who came in last place?
>
> $-4, -2, -1$
>
>   *Clark won the game, and Henry came in last place. I ordered the numbers on a number line, and -4 is farthest to the left. That means -4 is the smallest of the three numbers, so Henry lost. Next on the number line is -2, which is to the right of -4 and to the left of -1. Farthest to the right is -1, which is the greatest of the three negative numbers, so Clark won the game.*
>
> 6. Represent each of the following elevations using a rational number. Then, order the numbers from least to greatest.
>
>   Cayuga Lake   122 meters above sea level
>
>   Mount Marcy   1,629 meters above sea level
>
>   New York Stock Exchange Vault   15.24 meters below sea level
>
> $-15.24; 122; 1,629$
>
>   *I drew a number line, and -15.24 is the only number to the left of zero, so it is the least (because as you move to the right, the numbers increase). Next on the number line is 122, which is to the right of zero. Last on the number line is $1,629$, which is to the right of 122, so $1,629$ meters is the greatest elevation.*

Closing (5 minutes): What Is the Value of Each Number, and Which Is Larger?

This Closing is a verbal game similar to the Opening Exercise. However, instead of asking questions, students use the two clues the teacher provides and, as a visual model, the number line diagram in their materials. Using the clues and the number line, students determine the value of each number and which number is greater. When a student raises his hand and is called upon, he states the values and which is greater.

Scaffolding:
- *Some students need to rely on the number line for this activity. Others do not, as they have moved from the pictorial to the abstract stage.*
- *Adjust the wait time and difficulty level of the questions to meet the needs of learners.*

Listed below are five rounds of potential questions.

- **Round 1:** The first number is 3 units to the right of -1. The second number is 5 units to the left of 4. What is the value of each number, and which is greater?
    - *First Number: 2; Second Number: -1; 2 is greater than -1.*
- **Round 2:** The first number is neither positive nor negative. The second number is 1 unit to the left of -1. What is the value of each number, and which is greater?
    - *First Number: 0; Second Number: -2; 0 is greater than -2.*
- **Round 3:** The first number is $8\frac{1}{2}$ units to the right of -5. The second number is 3 units to the right of 0. What is the value of each number, and which is greater?
    - *First Number: 3.5; Second Number: 3; 3.5 is greater than 3.*

Lesson 7:   Ordering Integers and Other Rational Numbers

- **Round 4:** The first number is $\frac{1}{4}$ unit to the left of -7. The second number is 8 units to the left of 1. What is the value of each number, and which is greater?
  - *First Number: -7.25; Second Number: -7; -7 is greater than -7.25.*
- **Round 5:** The opposite of the first number is 2 units to the right of 3. The opposite of the second number is 2 units to the left of -3. What is the value of each number, and which is greater?
  - *First Number: -5; Second Number: 5; 5 is greater than -5.*

> **Closing:** What Is the Value of Each Number, and Which Is Larger?
>
> Use your teacher's verbal clues and this number line to determine which number is larger.
>
>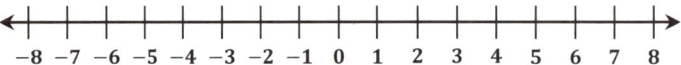

Exit Ticket (5 minutes)

Name _____ Date _____

Lesson 7: Ordering Integers and Other Rational Numbers

Exit Ticket

In math class, Christina and Brett are debating the relationship between two rational numbers. Read their claims below, and then write an explanation of who is correct. Use a number line model to support your answer.

<u>Christina's Claim</u>: "I know that 3 is greater than $2\frac{1}{2}$. So, -3 must be greater than $-2\frac{1}{2}$."

<u>Brett's Claim</u>: "Yes, 3 is greater than $2\frac{1}{2}$, but when you look at their opposites, their order will be opposite. So, that means $-2\frac{1}{2}$ is greater than -3."

Exit Ticket Sample Solutions

In math class, Christina and Brett are debating the relationship between two rational numbers. Read their claims below, and then write an explanation of who is correct. Use a number line model to support your answer.

<u>Christina's Claim</u>: "I know that 3 is greater than $2\frac{1}{2}$. So, -3 must be greater than $-2\frac{1}{2}$."

<u>Brett's Claim</u>: "Yes, 3 is greater than $2\frac{1}{2}$, but when you look at their opposites, their order will be opposite. So, that means $-2\frac{1}{2}$ is greater than -3."

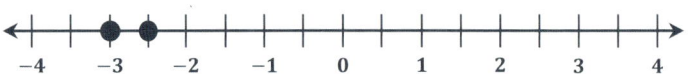

Brett is correct. I graphed the numbers on the number line, and -3 is to the left of $-2\frac{1}{2}$. The numbers increase as you move to the right, so $-2\frac{1}{2}$ is greater than -3.

Problem Set Sample Solutions

1. In the table below, list each set of rational numbers in order from least to greatest. Then, list their opposites. Finally, list the opposites in order from least to greatest. The first example has been completed for you.

Rational Numbers	Ordered from Least to Greatest	Opposites	Opposites Ordered from Least to Greatest
$-7.1, -7.25$	$-7.25, -7.1$	$7.25, 7.1$	$7.1, 7.25$
$\frac{1}{4}, -\frac{1}{2}$	$-\frac{1}{2}, \frac{1}{4}$	$\frac{1}{2}, -\frac{1}{4}$	$-\frac{1}{4}, \frac{1}{2}$
$2, -10$	$-10, 2$	$10, -2$	$-2, 10$
$0, 3\frac{1}{2}$	$0, 3\frac{1}{2}$	$0, -3\frac{1}{2}$	$-3\frac{1}{2}, 0$
$-5, -5.6$	$-5.6, -5$	$5.6, 5$	$5, 5.6$
$24\frac{1}{2}, 24$	$24, 24\frac{1}{2}$	$-24, -24\frac{1}{2}$	$-24\frac{1}{2}, -24$
$-99.9, -100$	$-100, -99.9$	$100, 99.9$	$99.9, 100$
$-0.05, -0.5$	$-0.5, -0.05$	$0.5, 0.05$	$0.05, 0.5$
$-0.7, 0$	$-0.7, 0$	$0.7, 0$	$0, 0.7$
$100.02, 100.04$	$100.02, 100.04$	$-100.02, -100.04$	$-100.04, -100.02$

2. For each row, what pattern do you notice between the numbers in the second and fourth columns? Why is this so?

For each row, the numbers in the second and fourth columns are opposites, and their order is opposite. This is because on the number line, as you move to the right, numbers increase. But as you move to the left, the numbers decrease. So, when comparing 5 and 10, 10 is to the right of 5; therefore, 10 is greater than 5. However, -10 is to the left of -5; therefore, -10 is less than -5.

A STORY OF RATIOS   Lesson 8   6•3

 Lesson 8: Ordering Integers and Other Rational Numbers

Student Outcomes

- Students write, interpret, and explain statements of order for rational numbers in the real world.
- Students recognize that if $a < b$, then $-a > -b$ because a number and its opposite are equal distances from zero, and moving along the horizontal number line to the right means the numbers are increasing.

Lesson Notes

As a continuation of Lesson 7, students order rational numbers from least to greatest and from greatest to least. They relate the orderings to numbers' locations on the number line.

Classwork

Opening Exercise (6 minutes)

For this warm-up exercise, students work in groups of three or four to order the following rational numbers from least to greatest. Each group of students may be provided with cards to put in order, or the numbers may be displayed on the board where students work at their seats, recording them in the correct order. As an alternative, the numbers may be displayed on an interactive board along with a number line, and students or teams come up to the board and slide the numbers onto the number line into the correct order. Allow time for the class to come to a consensus on the correct order and for students to share with the class their strategies and thought processes.

The following are examples of rational numbers to sort and order:

$0, -4, \frac{1}{4}, -\frac{1}{2}, 1, -3\frac{3}{5}, 2, -4.1, -0.6, \frac{23}{5}, 6, -1, 4.5, -5, 2.1$

Solution:

$-5, -4.1, -4, -3\frac{3}{5}, -1, -0.6, -\frac{1}{2}, 0, \frac{1}{4}, 1, 2, 2.1, 4.5, \frac{23}{5}, 6$

> *Scaffolding:*
> Adjust the number of cards given to students depending on their ability level. The types of rational numbers given to each group of students may also be differentiated.

The following line of questioning can be used to elicit student responses:

- How did you begin to sort and order the numbers? What was your first step?
  - *Our group began by separating the numbers into two groups: negative numbers and positive numbers. Zero was not in either group, but we knew it fell in between the negative numbers and positive numbers.*
- What was your next step? What did you do with the two groups of numbers?
  - *We ordered the positive whole numbers and then took the remaining positive numbers and determined which two whole numbers they fell in between.*

 Lesson 8:   Ordering Integers and Other Rational Numbers   71

- How did you know where to place $\frac{1}{4}$ and $\frac{23}{5}$?
  - Since $\frac{1}{4}$ is less than a whole (1) but greater than zero, we knew the rational number was located between 0 and 1.
  - We know that $\frac{23}{5}$ is the same as $4\frac{3}{5}$, which is more than 4 but less than 5, so we knew the rational number was located between 4 and 5.
- How did you order the negative numbers?
  - First, we started with the negative integers: $-5, -4,$ and -1. -5 is the least because it is farthest left at 5 units to the left of zero. Then came -4, and then came -1, which is only 1 unit to the left of zero.
- How did you order the negative non-integers?
  - We know $-\frac{1}{2}$ is equivalent to $-\frac{5}{10}$, which is to the right of -0.6 (or $-\frac{6}{10}$) since -0.5 is closer to zero than -0.6. Then, we ordered -4.1 and $-3\frac{3}{5}$. Both numbers are close to -4, but -4.1 is to the left of -4, and $-3\frac{3}{5}$ is to the right of -4 and to the left of -3. Lastly, we put our ordered group of negative numbers to the left of zero and our ordered group of positive numbers to the right of zero and ended up with

$$-5, -4.1, \ -4, -3\tfrac{3}{5}, -1, -0.6, -\tfrac{1}{2}, 0, \tfrac{1}{4}, 1, 2, 2.1, 4.5, \tfrac{23}{5}, 6.$$

Exercise 1 (8 minutes)

1. Students are each given four index cards or small slips of paper. Each student must independently choose four non-integer rational numbers and write each one on a slip of paper. At least two of the numbers must be negative.
2. Students order their rational numbers from least to greatest by sliding their slips of paper into the correct order. The teacher walks around the room to check for understanding and to provide individual assistance. Students may use the number line in their student materials to help determine the order.
3. Once all students have arranged their numbers into the correct order, they shuffle them and then switch with another student.
4. Students arrange the new set of cards they receive into the correct order from least to greatest.
5. The pairs of students who exchanged cards discuss their solutions and come to a consensus.

Example 1 (3 minutes): Ordering Rational Numbers from Least to Greatest

Example 1: Ordering Rational Numbers from Least to Greatest

Sam has 10.00 in the bank. He owes his friend Hank 2.25. He owes his sister 1.75. Consider the three rational numbers related to this story of Sam's money. Write and order them from least to greatest.

$-2.25, -1.75, 10.00$

Scaffolding:

Provide a number line diagram for visual learners to help them determine the numbers' order.

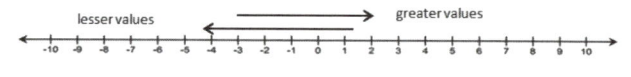

- Explain the process you used to determine the order of the numbers.
  - There is only one positive number, 10.00, so I know that 10.00 is the greatest. I know 2.25 is farther to the right on the number line than 1.75; therefore, its opposite, −2.25, will be farther to the left than the opposite of 1.75. This means −2.25 is the least, and −1.75 is between −2.25 and 10.00.
- How would the order change if you were asked to write the numbers from greatest to least?
  - The order would be reversed. I would list the numbers so that the number that comes first is the one farthest to the right on the number line, and the number that comes last is the one farthest to the left on the number line. The order would be 10.00 (the greatest), followed by −1.75, and then followed by −2.25 (the least).

Exercises 2–4 (10 minutes)

Allow time for students to share their answers with the class and explain their reasoning.

Exercises 2–4

For each problem, list the rational numbers that relate to each situation. Then, order them from least to greatest, and explain how you made your determination.

2. During their most recent visit to the optometrist (eye doctor), Kadijsha and her sister, Beth, had their vision tested. Kadijsha's vision in her left eye was −1.50, and her vision in her right eye was the opposite number. Beth's vision was −1.00 in her left eye and +0.25 in her right eye.

   −1.50, −1.00, 0.25, 1.50

   The opposite of −1.50 is 1.50, and 1.50 is farthest right on the number line, so it is the greatest. −1.50 is the same distance from zero but on the other side, so it is the least number. −1.00 is to the right of −1.50, so it is greater than −1.50, and 0.25 is to the right of −1.00, so it is greater than −1.00. Finally, 1.50 is the greatest.

3. There are three pieces of mail in Ms. Thomas's mailbox: a bill from the phone company for $38.12, a bill from the electric company for $67.55, and a tax refund check for $25.89. (A bill is money that you owe, and a tax refund check is money that you receive.)

   −67.55, −38.12, 25.89

   The change in Ms. Thomas's money is represented by −38.12 due to the phone bill, and −67.55 represents the change in her money due to the electric bill. Since −67.55 is farthest to the left on the number line, it is the least. Since −38.12 is to the right of −67.55, it comes next. The check she has to deposit for $25.89 can be represented by 25.89, which is to the right of −38.12, and so it is the greatest number.

4. Monica, Jack, and Destiny measured their arm lengths for an experiment in science class. They compared their arm lengths to a standard length of 22 inches. The listing below shows, in inches, how each student's arm length compares to 22 inches.

   Monica: $-\frac{1}{8}$

   Jack: $1\frac{3}{4}$

   Destiny: $-\frac{1}{2}$

   $-\frac{1}{2}, -\frac{1}{8}, 1\frac{3}{4}$

   I ordered the numbers on a number line, and $-\frac{1}{2}$ was farthest to the left. To the right of that was $-\frac{1}{8}$. Lastly, $1\frac{3}{4}$ is to the right of $-\frac{1}{8}$, so $1\frac{3}{4}$ is the greatest.

Lesson 8: Ordering Integers and Other Rational Numbers

A STORY OF RATIOS   Lesson 8   6•3

Example 2 (3 minutes): Ordering Rational Numbers from Greatest to Least

> **Example 2: Ordering Rational Numbers from Greatest to Least**
>
> Jason is entering college and has opened a checking account, which he will use for college expenses. His parents gave him 200.00 to deposit into the account. Jason wrote a check for 85.00 to pay for his calculus book and a check for 25.34 to pay for miscellaneous school supplies. Write the three rational numbers related to the balance in Jason's checking account in order from greatest to least.
>
> $200.00, -25.34, -85.00$

- Explain the process you used to determine the order of the numbers.
    - *There was only one positive number, 200.00, so I know that 200.00 is the greatest. I know 85.00 is farther to the right on the number line than 25.34, so its opposite, -85.00, will be farther to the left than the opposite of 25.34. This means -85.00 is the least, and -25.34 would be between -85.00 and 200.00.*

Exercises 5–6 (6 minutes)

Allow time for students to share their answers with the class and explain their reasoning.

> **Exercises 5–6**
>
> For each problem, list the rational numbers that relate to each situation in order from greatest to least. Explain how you arrived at the order.
>
> 5. The following are the current monthly bills that Mr. McGraw must pay:
> 122.00 Cable and Internet
> 73.45 Gas and Electric
> 45.00 Cell Phone
>
> $-45.00, -73.45, -122.00$
>
>   *Because Mr. McGraw owes the money, I represented the amount of each bill as a negative number. Ordering them from greatest to least means I have to move from right to left on a number line. Since -45.00 is farthest right, it is the greatest. To the left of that is -73.45, and to the left of that is -122.00, which means -122.00 is the least.*
>
> 6. $-\frac{1}{3}, 0, -\frac{1}{5}, \frac{1}{8}$
>
> $\frac{1}{8}, 0, -\frac{1}{5}, -\frac{1}{3}$
>
>   *I graphed them on the number line. Since I needed to order them from greatest to least, I moved from right to left to record the order. Farthest to the right is $\frac{1}{8}$, so that is the greatest value. To the left of that number is 0. To the left of 0 is $-\frac{1}{5}$, and the farthest left is $-\frac{1}{3}$, so that is the least.*

74 Lesson 8: Ordering Integers and Other Rational Numbers

Closing (3 minutes)

- If three numbers are ordered from least to greatest and the order is a, b, c, what would the order be if the same three numbers were arranged in order from greatest to least? How did you determine the new order?
  - c, b, a
  - *This is the correct order because it has to be exactly the opposite order since we are now moving right to left on the number line, when originally we moved left to right.*
- How does graphing numbers on a number line help us determine the order when arranging the numbers from greatest to least or least to greatest?
  - *Using a number line helps us order numbers because when numbers are placed on a number line, they are placed in order.*

Lesson Summary

When we order rational numbers, their opposites are in the opposite order. For example, if 7 is greater than 5, -7 is less than -5.

Exit Ticket (6 minutes)

Name _____ Date _____

Lesson 8: Ordering Integers and Other Rational Numbers

Exit Ticket

Order the following set of rational numbers from least to greatest, and explain how you determined the order.

$$-3,\ 0,\ -\tfrac{1}{2},\ 1,\ -3\tfrac{1}{3},\ 6,\ 5,\ -1,\ \tfrac{21}{5},\ 4$$

A STORY OF RATIOS Lesson 8 6•3

Exit Ticket Sample Solutions

Order the following set of rational numbers from least to greatest, and explain how you determined the order.

$$-3, 0, -\frac{1}{2}, 1, -3\frac{1}{3}, 6, 5, -1, \frac{21}{5}, 4$$

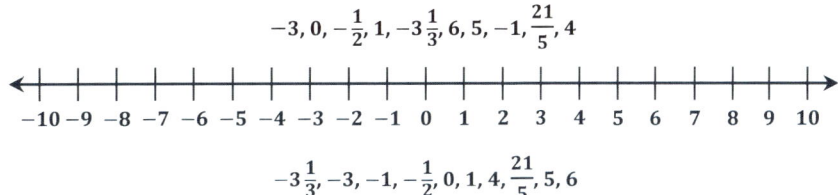

$$-3\frac{1}{3}, -3, -1, -\frac{1}{2}, 0, 1, 4, \frac{21}{5}, 5, 6$$

I drew a number line and started at zero. I located the positive numbers to the right and their opposites (the negative numbers) to the left of zero. The positive integers listed in order from left to right are 1, 4, 5, 6. And since $\frac{21}{5}$ is equal to $4\frac{1}{5}$, I know that it is $\frac{1}{5}$ more than 4 but less than 5. Therefore, I arrived at 0, 1, 4, $\frac{21}{5}$, 5, 6. Next, I ordered the negative numbers. Since -1 and -3 are the opposites of 1 and 3, they are 1 unit and 3 units from zero but to the left of zero. And $-3\frac{1}{3}$ is even farther left, since it is $3\frac{1}{3}$ units to the left of zero. The smallest number is farthest to the left, so I arrived at the following order: $-3\frac{1}{3}, -3, -1, -\frac{1}{2}, 0, 1, 4, \frac{21}{5}, 5, 6$.

Problem Set Sample Solutions

1.
   a. In the table below, list each set of rational numbers from greatest to least. Then, in the appropriate column, state which number was farthest right and which number was farthest left on the number line.

Column 1	Column 2	Column 3	Column 4
Rational Numbers	Ordered from Greatest to Least	Farthest Right on the Number Line	Farthest Left on the Number Line
$-1.75, -3.25$	$-1.75, -3.25$	-1.75	-3.25
$-9.7, -9$	$-9, -9.7$	-9	-9.7
$\frac{4}{5}, 0$	$\frac{4}{5}, 0$	$\frac{4}{5}$	0
$-70, -70\frac{4}{5}$	$-70, -70\frac{4}{5}$	-70	$-70\frac{4}{5}$
$-15, -5$	$-5, -15$	-5	-15
$\frac{1}{2}, -2$	$\frac{1}{2}, -2$	$\frac{1}{2}$	-2
$-99, -100, -99.3$	$-99, -99.3, -100$	-99	-100
$0.05, 0.5$	$0.5, 0.05$	0.5	0.05
$0, -\frac{3}{4}, -\frac{1}{4}$	$0, -\frac{1}{4}, -\frac{3}{4}$	0	$-\frac{3}{4}$
$-0.02, -0.04$	$-0.02, -0.04$	-0.02	-0.04

Lesson 8: Ordering Integers and Other Rational Numbers

b. For each row, describe the relationship between the number in Column 3 and its order in Column 2. Why is this?

The number in Column 3 is the first number listed in Column 2. Since it is farthest right on the number line, it will be the greatest; therefore, it comes first when ordering the numbers from greatest to least.

c. For each row, describe the relationship between the number in Column 4 and its order in Column 2. Why is this?

The number in Column 4 is the last number listed in Column 2. Since it is farthest left on the number line, it will be the smallest; therefore, it comes last when ordering the numbers from greatest to least.

2. If two rational numbers, a and b, are ordered such that a is less than b, then what must be true about the order for their opposites: $-a$ and $-b$?

The order will be reversed for the opposites, which means $-a$ is greater than $-b$.

3. Read each statement, and then write a statement relating the *opposites* of each of the given numbers:
   a. 7 is greater than 6.

   -7 *is less than* -6.

   b. 39.2 is greater than 30.

   -39.2 *is less than* -30.

   c. $-\frac{1}{5}$ is less than $\frac{1}{3}$.

   $\frac{1}{5}$ *is greater than* $-\frac{1}{3}$.

4. Order the following from least to greatest: -8, -19, 0, $\frac{1}{2}$, $\frac{1}{4}$.

   $-19, -8, 0, \frac{1}{4}, \frac{1}{2}$

5. Order the following from greatest to least: -12, 12, -19, $1\frac{1}{2}$, 5.

   $12, 5, 1\frac{1}{2}, -12, -19$

A STORY OF RATIOS   Lesson 9   6•3

 Lesson 9: Comparing Integers and Other Rational Numbers

Student Outcomes

- Students compare and interpret rational numbers' order on the number line, making statements that relate the numbers' location on the number line to their order.
- Students apply their prerequisite knowledge of place value, decimals, and fractions to compare integers and other rational numbers.
- Students relate integers and other rational numbers to real-world situations and problems.

Lesson Notes

Students complete an activity during Example 2 that requires preparation. The Activity Cards (attached to the end of the lesson) need to be prepared before the delivery of this lesson.

Classwork

Example 1 (3 minutes): Interpreting Number Line Models to Compare Numbers

Refer to the number line diagram below, which is also located in the student materials. In a whole-group discussion, create a real-world situation that relates to the numbers graphed on the number line. Include an explanation of what zero represents. Students should contribute suggestions to help the story evolve and come to a final state. Students write the related story in their student materials.

> **Example 1: Interpreting Number Line Models to Compare Numbers**
>
>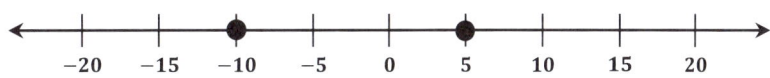
>
> *Answers may vary. Every August, the Boy Scouts go on an 8-day 40-mile hike. At the halfway point (20 miles into the hike), there is a check-in station for Scouts to check in and register. Thomas and Evan are Scouts in 2 different hiking groups. By Wednesday morning, Evan's group has 10 miles to go before it reaches the check-in station, and Thomas's group is 5 miles beyond the station. Zero on the number line represents the check-in station.*

 Lesson 9:   Comparing Integers and Other Rational Numbers   79

This work is derived from Eureka Math ™ and licensed by Great Minds. ©2015 Great Minds. eureka-math.org
G6-M3-TE-B3-1.3.0-07.2015

A STORY OF RATIOS   Lesson 9   6•3

Exercise 1 (7 minutes)

Display the following vertical number line model on the board. Students are to independently interpret the number line model to describe a real-world situation involving these two rational numbers. Remind students to compare the numbers and describe their order in their write-ups. After allowing adequate time for students to write their solutions, several students share what they wrote with the class. Students in the class determine whether the written responses correctly relate to the number line models.

Scaffolding:
Provide a story starter for students who are struggling to begin the writing task.

Exercise 1

1. Create a real-world situation that relates to the points shown in the number line model. Be sure to describe the relationship between the values of the two points and how it relates to their order on the number line.

   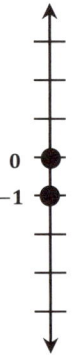

   Answers will vary.

   Alvin lives in Canada and is recording the outside temperature each night before he goes to bed. On Monday night, he recorded a temperature of 0 degrees Celsius. On Tuesday night, he recorded a temperature of −1 degree Celsius. Tuesday night's temperature was colder than Monday night's temperature. −1 is less than 0, so the associated point is below 0 on a vertical number line.

Example 2 (10 minutes)

Students are seated in groups of three or four, and each group is given a set of Activity Cards. For each group, photocopy, cut out, and scramble both sheets of Activity Cards that appear at the end of this lesson.

- Each group of students matches each word story card to its related number line card.
- For each number line diagram, students must write a statement relating the numbers' placement on the number line to their order.
- If time permits, the class goes over each answer as a whole group. For each number line diagram, students present their written statements as verbal statements to the class.
- An example follows:

| The Navy Seals are practicing new techniques. The blue submarine is 450 ft. below sea level, while the red submarine is 375 ft. below sea level. | 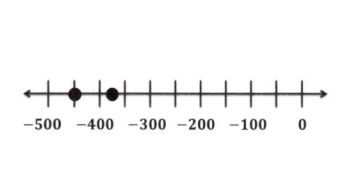 |

  □ *The blue submarine is farther below sea level than the red submarine because −450 is to the left of −375 on the number line; it is less than −375.*

A STORY OF RATIOS   Lesson 9   6•3

Exercises 2–8 (15 minutes)

Students read each of the following scenarios and decide whether they agree or disagree. They must defend and explain their stance in writing. Allow time for students to share their answers with the class and explain their reasoning. The class should come to a consensus for each one.

Scaffolding:
Provide a set of horizontal and vertical number lines for visual learners to create a number line model for each exercise.

MP.2 & MP.3

Exercises 2–8

For each problem, determine if you *agree* or *disagree* with the representation. Then, defend your stance by citing specific details in your writing.

2. Felicia needs to write a story problem that relates to the order in which the numbers $-6\frac{1}{2}$ and -10 are represented on a number line. She writes the following:

   "During a recent football game, our team lost yards on two consecutive downs. We lost $6\frac{1}{2}$ yards on the first down. During the second down, our quarterback was sacked for an additional 10-yard loss. On the number line, I represented this situation by first locating $-6\frac{1}{2}$. I located the point by moving $6\frac{1}{2}$ units to the left of zero. Then, I graphed the second point by moving 10 units to the left of 0."

   Agree. -10 is less than $-6\frac{1}{2}$ since -10 is to the left of $-6\frac{1}{2}$ on the number line. Since both numbers are negative, they indicate the team lost yards on both football plays, but they lost more yards on the second play.

3. Manuel looks at a number line diagram that has the points $-\frac{3}{4}$ and $-\frac{1}{2}$ graphed. He writes the following related story:

   "I borrowed 50 cents from my friend, Lester. I borrowed 75 cents from my friend, Calvin. I owe Lester less than I owe Calvin."

   Agree. $-\frac{3}{4}$ is equivalent to -0.75 and $-\frac{1}{2}$ is equivalent to -0.50. -0.50 and -0.75 both show that he owes money. But -0.50 is farther to the right on a number line, so Manuel does not owe Lester as much as he owes Calvin.

4. Henry located $2\frac{1}{4}$ and 2.1 on a number line. He wrote the following related story:

   "In gym class, both Jerry and I ran for 20 minutes. Jerry ran $2\frac{1}{4}$ miles, and I ran 2.1 miles. I ran a farther distance."

   Disagree. $2\frac{1}{4}$ is greater than 2.1 since $2\frac{1}{4}$ is equivalent to 2.25. On the number line, the point associated with 2.25 is to the right of 2.1. Jerry ran a farther distance.

5. Sam looked at two points that were graphed on a vertical number line. He saw the points -2 and 1.5. He wrote the following description:

   "I am looking at a vertical number line that shows the location of two specific points. The first point is a negative number, so it is below zero. The second point is a positive number, so it is above zero. The negative number is -2. The positive number is $\frac{1}{2}$ unit more than the negative number."

   Disagree. Sam was right when he said the negative number is below zero and the positive number is above zero. But 1.5 is $1\frac{1}{2}$ units above zero, and -2 is 2 units below zero. So, altogether, that means the positive number is $3\frac{1}{2}$ units more than -2.

Lesson 9:   Comparing Integers and Other Rational Numbers   81

A STORY OF RATIOS   Lesson 9   6•3

MP.2 & MP.3

6. Claire draws a vertical number line diagram and graphs two points: -10 and 10. She writes the following related story:

   "These two locations represent different elevations. One location is 10 feet above sea level, and one location is 10 feet below sea level. On a number line, 10 feet above sea level is represented by graphing a point at 10, and 10 feet below sea level is represented by graphing a point at -10."

   Agree. Zero in this case represents sea level. Both locations are 10 feet from zero but in opposite directions, so they are graphed on the number line at 10 and -10.

7. Mrs. Kimble, the sixth-grade math teacher, asked the class to describe the relationship between two points on the number line, 7.45 and 7.5, and to create a real-world scenario. Jackson writes the following story:

   "Two friends, Jackie and Jennie, each brought money to the fair. Jackie brought more than Jennie. Jackie brought $\$7.45$, and Jennie brought $\$7.50$. Since 7.45 has more digits than 7.5, it would come after 7.5 on the number line, or to the right, so it is a greater value."

   Disagree. Jackson is wrong by saying that 7.45 is to the right of 7.5 on the number line. 7.5 is the same as 7.50, and it is greater than 7.45. When I count by hundredths starting at 7.45, I would say $7.46, 7.47, 7.48, 7.49$, and then 7.50. So, 7.50 is greater than 7.45, and the associated point falls to the right of the point associated with 7.45 on the number line.

8. Justine graphs the points associated with the following numbers on a vertical number line: $-1\frac{1}{4}, -1\frac{1}{2}$, and 1. She then writes the following real-world scenario:

   "The nurse measured the height of three sixth-grade students and compared their heights to the height of a typical sixth grader. Two of the students' heights are below the typical height, and one is above the typical height. The point whose coordinate is 1 represents the student who has a height that is 1 inch above the typical height. Given this information, Justine determined that the student represented by the point associated with $-1\frac{1}{4}$ is the shortest of the three students."

   Disagree. Justine was wrong when she said the point $-1\frac{1}{4}$ represents the shortest of the three students. If zero stands for no change from the typical height, then the point associated with $-1\frac{1}{2}$ is farther below zero than the point associated with $-1\frac{1}{4}$. The greatest value is positive 1. Positive 1 represents the tallest person. The shortest person is represented by $-1\frac{1}{2}$.

Closing (4 minutes)

- How can use you use the number line to order a set of numbers? Will graphing the numbers on a vertical number line rather than a horizontal number line change this process?
    - *You can locate and graph the numbers on the number line to determine their order. If you use a vertical number line, their order is the same as it is on a horizontal number line, but instead of moving from left to right to go from least to greatest, you move from bottom to top. To determine the order of a set of numbers, the number that is farthest left (or farthest down on a vertical number line) is the smallest value. As you move right (or toward the top on a vertical number line), the numbers increase in value. So, the greatest number is graphed farthest right on a number line (or the highest one on a vertical number line).*

- If two points are graphed on a number line, what can you say about the value of the number associated with the point on the right in comparison to the value of the number associated with the point on the left?
    - *The number associated with the point on the right is greater than the number associated with the point on the left.*

82   Lesson 9:   Comparing Integers and Other Rational Numbers

This work is derived from Eureka Math ™ and licensed by Great Minds. ©2015 Great Minds. eureka-math.org
G6-M3-TE-B3-1.3.0-07.2015

- Which number is larger: -3.4 or $-3\frac{1}{2}$? How will graphing these numbers on a number line help you make this determination?
    - *Whichever number is graphed farthest to the left (or below) is the smaller number. In this example, $-3\frac{1}{2}$ would be graphed to the left of -3.4, so it is the smaller number. You can compare the numbers to make sure they are graphed correctly by either representing them both as a decimal or both as a fraction. $-3\frac{1}{2}$ is halfway between -3 and -4. So, if I divide the space into tenths, the associated point would be at -3.5 since $-3\frac{1}{2} = -3\frac{5}{10}$. When I graph -3.4, it would be 0.1 closer to -3, so it would be to the right of $-3\frac{1}{2}$. This means -3.4 is larger than $-3\frac{1}{2}$.*

Exit Ticket (6 minutes)

A STORY OF RATIOS　　　　　　　　　　　　　　　　　　　　　　Lesson 9　6•3

Name _____   Date _____

Lesson 9: Comparing Integers and Other Rational Numbers

Exit Ticket

1. Interpret the number line diagram shown below, and write a statement about the temperature for Tuesday compared to Monday at 11:00 p.m.

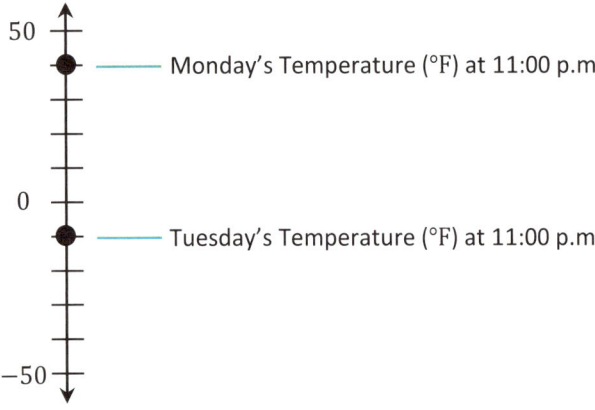

2. If the temperature at 11:00 p.m. on Wednesday is warmer than Tuesday's temperature but still below zero, what is a possible value for the temperature at 11:00 p.m. Wednesday?

Exit Ticket Sample Solutions

1. Interpret the number line diagram shown below, and write a statement about the temperature for Tuesday compared to Monday at 11:00 p.m.

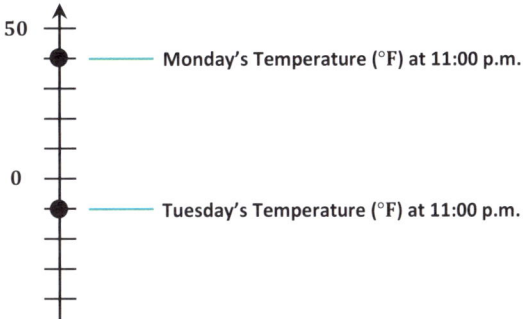

   At 11:00 p.m. on Monday, the temperature was about 40 degrees Fahrenheit, but at 11:00 p.m. on Tuesday, it was −10 degrees Fahrenheit. Tuesday's temperature of −10 degrees is below zero, but 40 degrees is above zero. It was much warmer on Monday at 11:00 p.m. than on Tuesday at that time.

2. If the temperature at 11:00 p.m. on Wednesday is warmer than Tuesday's temperature but still below zero, what is a possible value for the temperature at 11:00 p.m. Wednesday?

   Answers will vary but must be between 0 and −10. A possible temperature for Wednesday at 11:00 p.m. is −3 degrees Fahrenheit because −3 is less than zero and greater than −10.

Problem Set Sample Solutions

Write a story related to the points shown in each graph. Be sure to include a statement relating the numbers graphed on the number line to their order.

1.

   Answers will vary. Marcy earned no bonus points on her first math quiz. She earned 4 bonus points on her second math quiz. Zero represents earning no bonus points, and 4 represents earning 4 bonus points. Zero is graphed to the left of 4 on the number line. Zero is less than 4.

2.

   Answers will vary. My uncle's investment lost $200 in May. In June, the investment gained $150. The situation is represented by the points −200 and 150 on the vertical number line. Negative 200 is below zero, and 150 is above zero. −200 is less than 150.

Lesson 9: Comparing Integers and Other Rational Numbers

3.

    Answers will vary. I gave my sister $1.50 last week. This week, I gave her $0.50. The points -1.50 and -0.50 represent the change to my money supply. We know that -1.50 is to the left of -0.50 on the number line; therefore, -0.50 is greater than -1.50.

4.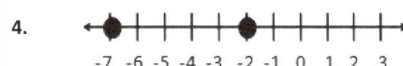

    Answers will vary. A fish is swimming 7 feet below the water's surface. A turtle is swimming 2 feet below the water's surface. We know that -7 is to the left of -2 on the number line. This means -7 is less than -2.

5.

    Answers will vary. I spent $8 on a CD last month. I earned $5 in allowance last month. -8 and 5 represent the changes to my money last month. -8 is to the left of 5 on a number line. -8 is 3 units farther away from zero than 5, which means that I spent $3 more on the CD than I made in allowance.

6.

    Answers will vary. Skip, Mark, and Angelo were standing in line in gym class. Skip was the third person behind Mark. Angelo was the first person ahead of Mark. If Mark represents zero on the number line, then Skip is associated with the point at -3, and Angelo is associated with the point at 1. 1 is 1 unit to the right of zero, and -3 is 3 units to the left of zero. -3 is less than 1.

7. 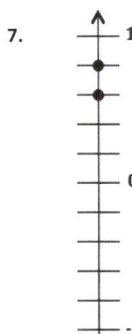 Answers will vary. I rode my bike $\frac{3}{5}$ miles on Saturday and $\frac{4}{5}$ miles on Sunday. On a vertical number line, $\frac{3}{5}$ and $\frac{4}{5}$ are both associated with points above zero, but $\frac{4}{5}$ is above $\frac{3}{5}$. This means that $\frac{4}{5}$ is greater than $\frac{3}{5}$.

Lesson 9

Activity Cards—Page 1

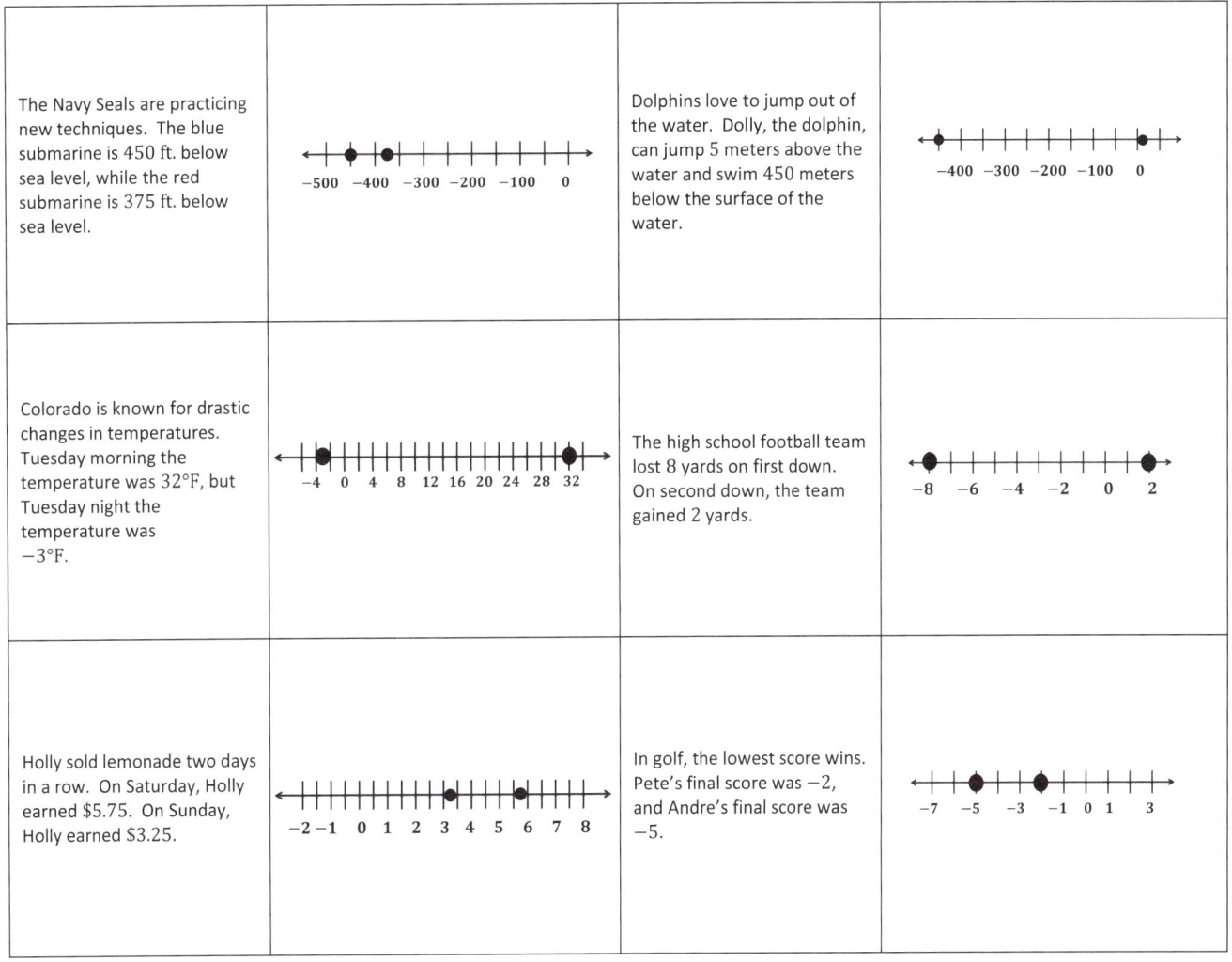

A STORY OF RATIOS Lesson 9 6•3

Activity Cards—Page 2

Scenario	Number Line	Scenario	Number Line
Teagon earned $450 last month cutting grass. Xavier spent $375 on a new computer.	Points at ~450 and −400 on a scale from −500 to 500	Jayden has earned 3 bonus points completing math extra credit assignments, while Shontelle has earned 32 bonus points.	Points at 32 and ~3 on a scale from 0 to 40
Kim and her friend Stacey went to the bookstore. Stacey spent $8 on notebooks. Kim spent $5 on snacks and pencils.	Points at −5 and −8 on a scale from 2 to −18	Last month, the stock market dropped $5\frac{3}{4}$ points overall. So far this month, the stock market rose $3\frac{1}{4}$ points.	Points at ~3¼ and ~−5¾ on a scale from 5 to −6
At a beach in California, if a person stands in the water, he is $\frac{1}{5}$ ft. below sea level. If the person walks onto the beach, he is $\frac{2}{5}$ ft. above sea level.	Points at 2/5 and −1/5 on a scale from 1 to −1	Brittany went to an office supply store twice last week. The first time, she made 2 copies that cost $0.20 each. The second time, she did not buy anything but found 2 dimes in the parking lot.	Points at ~0.20 and ~−0.40 on a scale from 1 to −1

88 Lesson 9: Comparing Integers and Other Rational Numbers

EUREKA MATH

Lesson 10: Writing and Interpreting Inequality Statements Involving Rational Numbers

Student Outcomes
- Students write and explain inequality statements involving rational numbers.
- Students justify inequality statements involving rational numbers.

Lesson Notes
Lessons 6, 7, 8, and 9 have prepared students for this lesson. Students have developed an understanding of how to represent, order, and compare rational numbers and now write and interpret inequality statements involving rational numbers. In Grade 5 Module 1, students wrote inequality statements involving decimals, and that experience serves as a foundation for this lesson as well.

Classwork

Opening Exercise (3 minutes)

As students enter the room, the following question is posted on the board (and is in the student materials).

"The amount of money I have in my pocket is less than $5 but greater than $4."

Direct students to discuss with their groups or elbow partners a possible value for the amount of money in your pocket, and then complete the following:

> **Opening Exercise**
>
> "The amount of money I have in my pocket is less than $5 but greater than $4."
>
> a. One possible value for the amount of money in my pocket is __$4.75__.
>
> b. Write an inequality statement comparing the possible value of the money in my pocket to $4.
>
> $4.00 < 4.75$
>
> c. Write an inequality statement comparing the possible value of the money in my pocket to $5.
>
> $4.75 < 5.00$

Scaffolding:
Use play money (including bills and coins) to represent $4 and $5. Students then create representations involving $4 and some change so as not to exceed $5.

MP.1

Discussion (5 minutes)

Allow time for students to share with the class their answers from part (a) of the Opening Exercise.

Discuss answers that would and would not fall between 4 and 5. Elicit student responses. Discuss potential answers that would or would not make sense in the context of the situation (e.g., 4.308 cannot be represented with physical money).

- What are some possible values for the amount of money in my pocket?
  - *$4.25, $4.39, $4.72, and $4.99*

Have several students write their inequality statements on the board from parts (b) and (c). Select a student's answer (or create another example of a correct answer) to use as a model. Write the answers to parts (b) and (c) side by side as shown below.

  - *4.00 < 4.75 and 4.75 < 5.00*

- Are there any integer solutions? Why or why not?
  - *There are no integer solutions because there are no integers between 4 and 5 because they are consecutive integers.*
- Are there any numbers between 4 and 5 that are not possible values for the amount of money in my pocket? Why or why not?
  - *We are talking about money, so all possible answers should be rational numbers that terminate at the hundredths place. There are more possible answers between 4 and 5, but they would not be accurate in this situation.*
- Is there a way to write one inequality statement that describes both of these relationships?
  - *Yes. 4.00 < 4.75 < 5.00*

Exercises 1–4 (4 minutes)

Students use a number line model to represent the order of the numbers used in their Opening Exercise (or in the example just discussed as a whole group). Students then graph three points: 4, 5, and a value that falls in between 4 and 5. Students should use the model and number line ordering to write one inequality statement relating the three numbers.

Exercises 1–4

1. Graph your answer from the Opening Exercise part (a) on the number line below.

2. Also, graph the points associated with 4 and 5 on the number line.

3. Explain in words how the location of the three numbers on the number line supports the inequality statements you wrote in the Opening Exercise parts (b) and (c).

   The numbers are ordered from least to greatest when I look at the number line from left to right. So, 4 is less than 4.75, and 4.75 is less than 5.

4. Write one inequality statement that shows the relationship among all three numbers.

   $4 < 4.75 < 5$

If students struggle with Exercise 4, spend adequate time as a whole group doing several examples of ordering students' sets of numbers using one statement of inequality.

The following two examples should be conducted as whole-group instruction.

A STORY OF RATIOS • Lesson 10 • 6•3

Example 1 (4 minutes): Writing Inequality Statements Involving Rational Numbers

Students should recall using inequality symbols to compare decimal numbers from Grade 5 Module 1. As needed, refer to the number line representation of non-integer rational numbers and their opposites from Lesson 6 of this module, and revisit the orientation of the less than symbol ($<$) and greater than symbol ($>$). Note that students may erroneously write $8 < 10\frac{1}{2} > 9$, which does not accurately describe the order of all three numbers.

Example 1: Writing Inequality Statements Involving Rational Numbers

Write one inequality statement to show the relationship among the following shoe sizes: $10\frac{1}{2}$, 8, and 9.

a. From least to greatest:

$$8 < 9 < 10\frac{1}{2}$$

b. From greatest to least:

$$10\frac{1}{2} > 9 > 8$$

Example 2 (4 minutes): Interpreting Data and Writing Inequality Statements

Example 2: Interpreting Data and Writing Inequality Statements

Mary is comparing the rainfall totals for May, June, and July. The data is reflected in the table below. Fill in the blanks below to create inequality statements that compare the Changes in Total Rainfall for each month (the right-most column of the table).

Month	This Year's Total Rainfall (in inches)	Last Year's Total Rainfall (in inches)	Change in Total Rainfall from Last Year to This Year (in inches)
May	2.3	3.7	−1.4
June	3.8	3.5	0.3
July	3.7	3.2	0.5

Write one inequality to order the Changes in Total Rainfall: $-1.4 < 0.3 < 0.5$ $0.5 > 0.3 > -1.4$

                                                             From least to greatest   From greatest to least

In this case, does the greatest number indicate the greatest change in rainfall? Explain.

No. In this situation, the greatest change is for the month of May since the average total rainfall went down from last year by 1.4 inches, but the greatest number in the inequality statement is 0.5.

Lesson 10: Writing and Interpreting Inequality Statements Involving Rational Numbers

Exercises 5–8 (8 minutes)

Students work independently to answer the following questions. Allow time for students to present their answers and share their thought processes to the class. Use the following as an optional task: Have students transfer their word problems for Exercise 8 onto paper using colorful markers or colored pencils, and display them in the classroom.

Exercises 5–8

5. Mark's favorite football team lost yards on two back-to-back plays. They lost 3 yards on the first play. They lost 1 yard on the second play. Write an inequality statement using integers to compare the forward progress made on each play.

   $-3 < -1$

6. Sierra had to pay the school for two textbooks that she lost. One textbook cost $55, and the other cost $75. Her mother wrote two separate checks, one for each expense. Write two integers that represent the change to her mother's checking account balance. Then, write an inequality statement that shows the relationship between these two numbers.

   -55 and -75; $-55 > -75$

7. Jason ordered the numbers -70, -18, and -18.5 from least to greatest by writing the following statement:
   $-18 < -18.5 < -70$.

   Is this a true statement? Explain.

   No, it is not a true statement because $18 < 18.5 < 70$, so the opposites of these numbers are in the opposite order. The order should be $-70 < -18.5 < 18$.

8. Write a real-world situation that is represented by the following inequality: $-19 < 40$. Explain the position of the numbers on a number line.

   The coldest temperature in January was -19 degrees Fahrenheit, and the warmest temperature was 40 degrees Fahrenheit. Since the point associated with 40 is above zero on a vertical number line and -19 is below zero, we know that 40 is greater than -19. This means that 40 degrees Fahrenheit is warmer than -19 degrees Fahrenheit.

Sprint (5 minutes): Writing Inequalities

Photocopy the attached two-page Sprints so that each student receives a copy. Time students, allowing one minute to complete Side A. Before students begin, inform them that they may not skip over questions and that they must move in order. After one minute, discuss the answers. Before administering Side B, elicit strategies from those students who were able to accurately complete many problems on Side A. Administer Side B in the same fashion, and review the answers. Refer to the Sprints and Sprint Delivery Script sections in the Module Overview for directions to administer a Sprint.

Exercise 9 (4 minutes): A Closer Look at the Sprint

Students are asked to look closely at two related examples from the Sprint and explain the relationship between the numbers' order, the inequality symbols, and the graphs of the numbers on the number line.

Exercise 9: A Closer Look at the Sprint

9. Look at the following two examples from the Sprint.

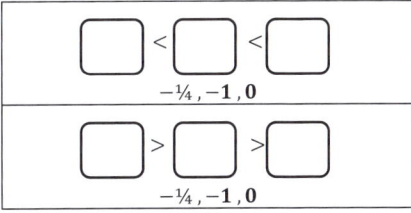

a. Fill in the numbers in the correct order.

$-1 < -\frac{1}{4} < 0$ and $0 > -\frac{1}{4} > -1$

b. Explain how the position of the numbers on the number line supports the inequality statements you created.

-1 is the farthest left on the number line, so it is the least value. 0 is farthest right, so it is the greatest value, and $-\frac{1}{4}$ is in between.

c. Create a new pair of greater than and less than inequality statements using three other rational numbers.

Answers will vary. $8 > 0.5 > -1.8$ and $-1.8 < 0.5 < 8$

Closing (3 minutes)

- What can you do before writing an inequality statement involving three numbers that makes it easier to write the inequality statement? For example, explain the process for writing one inequality statement comparing $-3, 8,$ and -10.
  - *First, I would order the numbers, either from least to greatest or greatest to least.*
- If you know the order of a set of numbers, how can you represent the order using inequality symbols?
  - *For example, if the numbers are 1, 2, and 3, you can either write $1 < 2 < 3$ or $3 > 2 > 1$.*
- If two negative numbers are ordered using the $<$ symbol, what must be true about their positions on a horizontal number line? On a vertical number line?
  - *The first number must be associated with a point to the left of the second number on a horizontal number line. The first number must be associated with a point below the second number on a vertical number line.*

Exit Ticket (5 minutes)

Lesson 10: Writing and Interpreting Inequality Statements Involving Rational Numbers

Name _____   Date _____

Lesson 10: Writing and Interpreting Inequality Statements Involving Rational Numbers

Exit Ticket

Kendra collected data for her science project. She surveyed people asking them how many hours they sleep during a typical night. The chart below shows how each person's response compares to 8 hours (which is the answer she expected most people to say).

Name	Number of Hours (usually slept each night)	Compared to 8 Hours
Frankie	8.5	0.5
Mr. Fields	7	−1.0
Karla	9.5	1.5
Louis	8	0
Tiffany	$7\frac{3}{4}$	$-\frac{1}{4}$

a. Plot and label each of the numbers in the right-most column of the table above on the number line below.

b. List the numbers from least to greatest.

c. Using your answer from part (b) and inequality symbols, write one statement that shows the relationship among all of the numbers.

A STORY OF RATIOS — Lesson 10 6•3

Exit Ticket Sample Solutions

Kendra collected data for her science project. She surveyed people asking them how many hours they sleep during a typical night. The chart below shows how each person's response compares to 8 hours (which is the answer she expected most people to say).

Name	Number of Hours (usually slept each night)	Compared to 8 Hours
Frankie	8.5	0.5
Mr. Fields	7	−1.0
Karla	9.5	1.5
Louis	8	0
Tiffany	$7\frac{3}{4}$	$-\frac{1}{4}$

a. Plot and label each of the numbers in the right-most column of the table above on the number line below.

b. List the numbers from least to greatest.

$-1.0,\ -\frac{1}{4},\ 0,\ 0.5,\ 1.5$

c. Using your answer from part (b) and inequality symbols, write one statement that shows the relationship among all the numbers.

$-1.0 < -\frac{1}{4} < 0 < 0.5 < 1.5$ or $1.5 > 0.5 > 0 > -\frac{1}{4} > -1.0$

Problem Set Sample Solutions

For each of the relationships described below, write an inequality that relates the rational numbers.

1. Seven feet below sea level is farther below sea level than $4\frac{1}{2}$ feet below sea level.

   $-7 < -4\frac{1}{2}$

2. Sixteen degrees Celsius is warmer than zero degrees Celsius.

   $16 > 0$

3. Three and one-half yards of fabric is less than five and one-half yards of fabric.

   $3\frac{1}{2} < 5\frac{1}{2}$

Lesson 10: Writing and Interpreting Inequality Statements Involving Rational Numbers

4. A loss of $500 in the stock market is worse than a gain of $200 in the stock market.

    $-500 < 200$

5. A test score of 64 is worse than a test score of 65, and a test score of 65 is worse than a test score of $67\frac{1}{2}$.

    $64 < 65 < 67\frac{1}{2}$

6. In December, the total snowfall was 13.2 inches, which is more than the total snowfall in October and November, which was 3.7 inches and 6.15 inches, respectively.

    $13.2 > 6.15 > 3.7$

For each of the following, use the information given by the inequality to describe the relative position of the numbers on a horizontal number line.

7. $-0.2 < -0.1$

    -0.2 is to the left of -0.1, or -0.1 is to the right of -0.2.

8. $8\frac{1}{4} > -8\frac{1}{4}$

    $8\frac{1}{4}$ is to the right of $-8\frac{1}{4}$, or $-8\frac{1}{4}$ is to the left of $8\frac{1}{4}$.

9. $-2 < 0 < 5$

    -2 is to the left of zero and zero is to the left of 5, or 5 is to the right of zero and zero is to the right of -2.

10. $-99 > -100$

    -99 is to the right of -100, or -100 is to the left of -99.

11. $-7.6 < -7\frac{1}{2} < -7$

    -7.6 is to the left of $-7\frac{1}{2}$ and $-7\frac{1}{2}$ is to the left of -7, or -7 is to the right of $-7\frac{1}{2}$ and $-7\frac{1}{2}$ is to the right of -7.6.

Fill in the blanks with numbers that correctly complete each of the statements.

12. Three integers between -4 and 0 $-3 < -2 < -1$

13. Three rational numbers between 16 and 15 $15.3 < 15.6 < 15.7$ Other answers are possible.

14. Three rational numbers between -1 and -2 $-1.9 < -1.55 < -1.02$ Other answers are possible.

15. Three integers between 2 and -2 $-1 < 0 < 1$

A STORY OF RATIOS Lesson 10 6•3

Number Correct: _____

Rational Numbers: Inequality Statements—Round 1

Directions: Work in numerical order to answer Problems 1–33. Arrange each set of numbers in order according to the inequality symbols.

#	Problem	#	Problem	#	Problem
1.	☐ < ☐ < ☐ 1, −1, 0	12.	☐ > ☐ > ☐ 7, −6, 6	23.	☐ > ☐ > ☐ 25, ¾, −¾
2.	☐ > ☐ > ☐ 1, −1, 0	13.	☐ > ☐ > ☐ 17, 4, 16	24.	☐ < ☐ < ☐ 25, ¾, −¾
3.	☐ < ☐ < ☐ 3 ½, −3 ½, 0	14.	☐ < ☐ < ☐ 17, 4, 16	25.	☐ > ☐ > ☐ 2.2, 2.3, 2.4
4.	☐ > ☐ > ☐ 3 ½, −3 ½, 0	15.	☐ < ☐ < ☐ 0, 12, −11	26.	☐ > ☐ > ☐ 1.2, 1.3, 1.4
5.	☐ > ☐ > ☐ 1, −½, ½	16.	☐ > ☐ > ☐ 0, 12, −11	27.	☐ > ☐ > ☐ 0.2, 0.3, 0.4
6.	☐ < ☐ < ☐ 1, −½, ½	17.	☐ > ☐ > ☐ 1, ¼, ½	28.	☐ > ☐ > ☐ −0.5, −1, −0.6
7.	☐ < ☐ < ☐ −3, −4, −5	18.	☐ < ☐ < ☐ 1, ¼, ½	29.	☐ < ☐ < ☐ −0.5, −1, −0.6
8.	☐ < ☐ < ☐ −13, −14, −15	19.	☐ < ☐ < ☐ −½, ½, 0	30.	☐ < ☐ < ☐ −8, −9, 8
9.	☐ > ☐ > ☐ −13, −14, −15	20.	☐ > ☐ > ☐ −½, ½, 0	31.	☐ < ☐ < ☐ −18, −19, −2
10.	☐ < ☐ < ☐ −¼, −1, 0	21.	☐ < ☐ < ☐ 50, −10, 0	32.	☐ > ☐ > ☐ −2, −3, 1
11.	☐ > ☐ > ☐ −¼, −1, 0	22.	☐ < ☐ < ☐ −50, 10, 0	33.	☐ < ☐ < ☐ −2, −3, 1

Lesson 10: Writing and Interpreting Inequality Statements Involving Rational Numbers

Rational Numbers: Inequality Statements—Round 1 [KEY]

Directions: Work in numerical order to answer Problems 1–33. Arrange each set of numbers in order according to the inequality symbols.

#	Inequality	Set
1.	$-1 < 0 < 1$	$1, -1, 0$
2.	$1 > 0 > -1$	$1, -1, 0$
3.	$-3\frac{1}{2} < 0 < 3\frac{1}{2}$	$3\frac{1}{2}, -3\frac{1}{2}, 0$
4.	$3\frac{1}{2} > 0 > -3\frac{1}{2}$	$3\frac{1}{2}, -3\frac{1}{2}, 0$
5.	$1 > \frac{1}{2} > -\frac{1}{2}$	$1, -\frac{1}{2}, \frac{1}{2}$
6.	$-\frac{1}{2} < \frac{1}{2} < 1$	$1, -\frac{1}{2}, \frac{1}{2}$
7.	$-5 < -4 < -3$	$-3, -4, -5$
8.	$-15 < -14 < -13$	$-13, -14, -15$
9.	$-13 > -14 > -15$	$-13, -14, -15$
10.	$-1 < -\frac{1}{4} < 0$	$-\frac{1}{4}, -1, 0$
11.	$0 > -\frac{1}{4} > -1$	$-\frac{1}{4}, -1, 0$
12.	$7 > 6 > -6$	$7, -6, 6$
13.	$17 > 16 > 4$	$17, 4, 16$
14.	$4 < 16 < 17$	$17, 4, 16$
15.	$-11 < 0 < 12$	$0, 12, -11$
16.	$12 > 0 > -11$	$0, 12, -11$
17.	$1 > \frac{1}{2} > \frac{1}{4}$	$1, \frac{1}{4}, \frac{1}{2}$
18.	$\frac{1}{4} < \frac{1}{2} < 1$	$1, \frac{1}{4}, \frac{1}{2}$
19.	$-\frac{1}{2} < 0 < \frac{1}{2}$	$-\frac{1}{2}, \frac{1}{2}, 0$
20.	$\frac{1}{2} > 0 > -\frac{1}{2}$	$-\frac{1}{2}, \frac{1}{2}, 0$
21.	$-10 < 0 < 50$	$50, -10, 0$
22.	$-50 < 0 < 10$	$-50, 10, 0$
23.	$25 > \frac{3}{4} > -\frac{3}{4}$	$25, \frac{3}{4}, -\frac{3}{4}$
24.	$-\frac{3}{4} < \frac{3}{4} < 25$	$25, \frac{3}{4}, -\frac{3}{4}$
25.	$2.4 > 2.3 > 2.2$	$2.2, 2.3, 2.4$
26.	$1.4 > 1.3 > 1.2$	$1.2, 1.3, 1.4$
27.	$0.4 > 0.3 > 0.2$	$0.2, 0.3, 0.4$
28.	$-0.5 > -0.6 > -1$	$-0.5, -1, -0.6$
29.	$-1 < -0.6 < -0.5$	$-0.5, -1, -0.6$
30.	$-9 < -8 < 8$	$-8, -9, 8$
31.	$-19 < -18 < -2$	$-18, -19, -2$
32.	$1 > -2 > -3$	$-2, -3, 1$
33.	$-3 < -2 < 1$	$-2, -3, 1$

A STORY OF RATIOS

Lesson 10 6•3

Number Correct: _____
Improvement: _____

Rational Numbers: Inequality Statements—Round 2

Directions: Work in numerical order to answer Problems 1–33. Arrange each set of numbers in order according to the inequality symbols.

#	Problem	#	Problem	#	Problem
1.	☐ < ☐ < ☐ $1/7, -1/7, 0$	12.	☐ > ☐ > ☐ $1\tfrac{1}{4}, 1, 1\tfrac{1}{2}$	23.	☐ > ☐ > ☐ $1, 1\tfrac{3}{4}, -1\tfrac{3}{4}$
2.	☐ > ☐ > ☐ $1/7, -1/7, 0$	13.	☐ > ☐ > ☐ $11\tfrac{1}{4}, 11, 11\tfrac{1}{2}$	24.	☐ < ☐ < ☐ $1, 1\tfrac{3}{4}, -1\tfrac{3}{4}$
3.	☐ < ☐ < ☐ $3/7, 2/7, -1/7$	14.	☐ < ☐ < ☐ $11\tfrac{1}{4}, 11, 11\tfrac{1}{2}$	25.	☐ > ☐ > ☐ $-82, -93, -104$
4.	☐ > ☐ > ☐ $3/7, 2/7, -1/7$	15.	☐ < ☐ < ☐ $0, 0.2, -0.1$	26.	☐ < ☐ < ☐ $-82, -93, -104$
5.	☐ > ☐ > ☐ $-4/5, 1/5, -1/5$	16.	☐ > ☐ > ☐ $0, 0.2, -0.1$	27.	☐ > ☐ > ☐ $0.5, 1, 0.6$
6.	☐ < ☐ < ☐ $-4/5, 1/5, -1/5$	17.	☐ > ☐ > ☐ $1, 0.7, 1/10$	28.	☐ > ☐ > ☐ $-0.5, -1, -0.6$
7.	☐ < ☐ < ☐ $-8/9, 5/9, 1/9$	18.	☐ < ☐ < ☐ $1, 0.7, 1/10$	29.	☐ < ☐ < ☐ $-0.5, -1, -0.6$
8.	☐ > ☐ > ☐ $-8/9, 5/9, 1/9$	19.	☐ < ☐ < ☐ $0, -12, -12\tfrac{1}{2}$	30.	☐ < ☐ < ☐ $1, 8, 9$
9.	☐ > ☐ > ☐ $-30, -10, -50$	20.	☐ > ☐ > ☐ $0, -12, -12\tfrac{1}{2}$	31.	☐ < ☐ < ☐ $-1, -8, -9$
10.	☐ < ☐ < ☐ $-30, -10, -50$	21.	☐ < ☐ < ☐ $5, -1, 0$	32.	☐ > ☐ > ☐ $-2, -3, -5$
11.	☐ > ☐ > ☐ $-40, -20, -60$	22.	☐ < ☐ < ☐ $-5, 1, 0$	33.	☐ > ☐ > ☐ $2, 3, 5$

Lesson 10: Writing and Interpreting Inequality Statements Involving Rational Numbers

Rational Numbers: Inequality Statements—Round 2 [KEY]

Directions: Work in numerical order to answer Problems 1–33. Arrange each set of numbers in order according to the inequality symbols.

#	Inequality	Given
1.	$-\frac{1}{7} < 0 < \frac{1}{7}$	$1/7, -1/7, 0$
2.	$\frac{1}{7} > 0 > -\frac{1}{7}$	$1/7, -1/7, 0$
3.	$-\frac{1}{7} < \frac{2}{7} < \frac{3}{7}$	$3/7, 2/7, -1/7$
4.	$\frac{3}{7} > \frac{2}{7} > -\frac{1}{7}$	$3/7, 2/7, -1/7$
5.	$\frac{1}{5} > -\frac{1}{5} > -\frac{4}{5}$	$-4/5, 1/5, -1/5$
6.	$-\frac{4}{5} < -\frac{1}{5} < \frac{1}{5}$	$-4/5, 1/5, -1/5$
7.	$-\frac{8}{9} < \frac{1}{9} < \frac{5}{9}$	$-8/9, 5/9, 1/9$
8.	$\frac{5}{9} > \frac{1}{9} > -\frac{8}{9}$	$-8/9, 5/9, 1/9$
9.	$-10 > -30 > -50$	$-30, -10, -50$
10.	$-50 < -30 < -10$	$-30, -10, -50$
11.	$-20 > -40 > -60$	$-40, -20, -60$
12.	$1\frac{1}{2} > 1\frac{1}{4} > 1$	$1\frac{1}{4}, 1, 1\frac{1}{2}$
13.	$11\frac{1}{2} > 11\frac{1}{4} > 11$	$11\frac{1}{4}, 11, 11\frac{1}{2}$
14.	$11 < 11\frac{1}{4} < 11\frac{1}{2}$	$11\frac{1}{4}, 11, 11\frac{1}{2}$
15.	$-0.1 < 0 < 0.2$	$0, 0.2, -0.1$
16.	$0.2 > 0 > -0.1$	$0, 0.2, -0.1$
17.	$1 > 0.7 > \frac{1}{10}$	$1, 0.7, 1/10$
18.	$\frac{1}{10} < 0.7 < 1$	$1, 0.7, 1/10$
19.	$-12\frac{1}{2} < -12 < 0$	$0, -12, -12\frac{1}{2}$
20.	$0 > -12 > -12\frac{1}{2}$	$0, -12, -12\frac{1}{2}$
21.	$-1 < 0 < 5$	$5, -1, 0$
22.	$-5 < 0 < 1$	$-5, 1, 0$
23.	$1\frac{3}{4} > 1 > -1\frac{3}{4}$	$1, 1\frac{3}{4}, -1\frac{3}{4}$
24.	$-1\frac{3}{4} < 1 < 1\frac{3}{4}$	$1, 1\frac{3}{4}, -1\frac{3}{4}$
25.	$-82 > -93 > -104$	$-82, -93, -104$
26.	$-104 < -93 < -82$	$-82, -93, -104$
27.	$1 > 0.6 > 0.5$	$0.5, 1, 0.6$
28.	$-0.5 > -0.6 > -1$	$-0.5, -1, -0.6$
29.	$-1 < -0.6 < -0.5$	$-0.5, -1, -0.6$
30.	$1 < 8 < 9$	$1, 8, 9$
31.	$-9 < -8 < -1$	$-1, -8, -9$
32.	$-2 > -3 > -5$	$-2, -3, -5$
33.	$5 > 3 > 2$	$2, 3, 5$

Lesson 10: Writing and Interpreting Inequality Statements Involving Rational Numbers

A STORY OF RATIOS   Lesson 11   6•3

 Lesson 11: Absolute Value—Magnitude and Distance

Student Outcomes

- Students understand the absolute value of a number as its distance from zero on the number line.
- Students use absolute value to find the magnitude of a positive or negative quantity in a real-world situation.

Classwork

Opening Exercise (4 minutes)

For this warm-up exercise, students work individually to record two different rational numbers that are the same distance from zero. Students find as many examples as possible and reach a conclusion about what must be true for every pair of numbers that lie that same distance from zero.

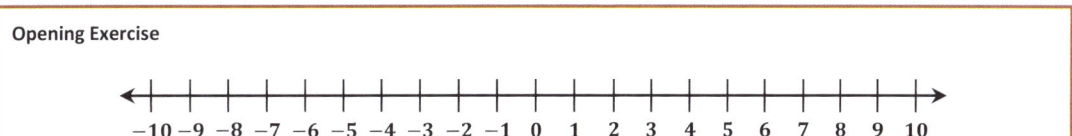
Opening Exercise

MP.8

After two minutes:

- What are some examples you found (pairs of numbers that are the same distance from zero)?
    - $-\frac{1}{2}$ and $\frac{1}{2}$, 8.01 and -8.01, -7 and 7.
- What is the relationship between each pair of numbers?
    - *They are opposites.*
- How does each pair of numbers relate to zero?
    - *Both numbers in each pair are the same distance from zero.*

Discussion (3 minutes)

- We just saw that every number and its opposite are the same distance from zero on the number line. The *absolute value* of a number is the distance between the number and zero on a number line.
- In other words, a number and its opposite have the same *absolute value*.
- What is the absolute value of 5? Explain.
    - *The absolute value of* 5 *is* 5 *because it is* 5 *units from zero.*
- What is the absolute value of -5?
    - *The absolute value of* -5 *is also* 5 *because it is also* 5 *units from zero.*
- Both 5 and -5 are five units from zero, which makes 5 and -5 opposites.

Scaffolding:

Provide students with a number line so they can physically count the number of units between a number and zero.

 Lesson 11:   Absolute Value—Magnitude and Distance   101

A STORY OF RATIOS

Lesson 11 6•3

- What is the absolute value of -1?
  - 1
- What other number has an absolute value of 1? Explain.
  - 1 *also has an absolute value of* 1 *because* 1 *and* -1 *are opposites, so they have the same absolute value.*
- What is the absolute value of 0?
  - 0

Example 1 (3 minutes): The Absolute Value of a Number

Example 1: The Absolute Value of a Number

The absolute value of ten is written as $|10|$. On the number line, count the number of units from 10 to 0. How many units is 10 from 0?

$|10| = 10$

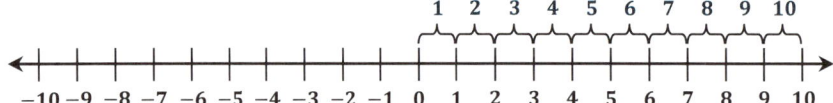

What other number has an absolute value of 10? Why?

$|-10| = 10$ *because* -10 *is* 10 *units from zero and* -10 *and* 10 *are opposites.*

The <u>absolute value</u> of a number is the distance between the number and zero on the number line.

Exercises 1–3 (4 minutes)

Exercises 1–3

Complete the following chart.

	Number	Absolute Value	Number Line Diagram	Different Number with the Same Absolute Value
1.	-6	$|-6| = 6$	←—+—→ $-10\,-9\,-8\,-7\,-6\,-5\,-4\,-3\,-2\,-1\;\;0\;\;1\;\;2\;\;3\;\;4\;\;5\;\;6\;\;7\;\;8\;\;9\;\;10$	6
2.	8	$|8| = 8$	←—+—→ $-10\,-9\,-8\,-7\,-6\,-5\,-4\,-3\,-2\,-1\;\;0\;\;1\;\;2\;\;3\;\;4\;\;5\;\;6\;\;7\;\;8\;\;9\;\;10$	-8
3.	-1	$|-1| = 1$	←—+—→ $-10\,-9\,-8\,-7\,-6\,-5\,-4\,-3\,-2\,-1\;\;0\;\;1\;\;2\;\;3\;\;4\;\;5\;\;6\;\;7\;\;8\;\;9\;\;10$	1

Lesson 11: Absolute Value—Magnitude and Distance

A STORY OF RATIOS   Lesson 11   6•3

Example 2 (3 minutes): Using Absolute Value to Find Magnitude

Example 2: Using Absolute Value to Find Magnitude

Mrs. Owens received a call from her bank because she had a checkbook balance of $-\$45$. What was the magnitude of the amount overdrawn?

$|-45| = 45$ *Mrs. Owens overdrew her checking account by $45.*

The <u>magnitude</u> of a measurement is the absolute value of its measure.

Exercises 4–8 (6 minutes)

Exercises 4–8

For each scenario below, use absolute value to determine the magnitude of each quantity.

4. Maria was sick with the flu, and her weight change as a result of it is represented by -4 pounds. How much weight did Maria lose?

   $|-4| = 4$ *Maria lost 4 pounds.*

5. Jeffrey owes his friend $5. How much is Jeffrey's debt?

   $|-5| = 5$ *Jeffrey has a $5 debt.*

6. The elevation of Niagara Falls, which is located between Lake Erie and Lake Ontario, is 326 feet. How far is this above sea level?

   $|326| = 326$ *It is 326 feet above sea level.*

7. How far below zero is -16 degrees Celsius?

   $|-16| = 16$ *$-16°C$ is 16 degrees below zero.*

8. Frank received a monthly statement for his college savings account. It listed a deposit of $100 as $+100.00$. It listed a withdrawal of $25 as -25.00. The statement showed an overall ending balance of $835.50. How much money did Frank add to his account that month? How much did he take out? What is the total amount Frank has saved for college?

   $|100| = 100$ *Frank added $100 to his account.*
   $|-25| = 25$ *Frank took $25 out of his account.*
   $|835.50| = 835.50$ *The total amount of Frank's savings for college is $835.50.*

MP.6

Lesson 11: Absolute Value—Magnitude and Distance   103

A STORY OF RATIOS   Lesson 11   6•3

Exercises 9–19 (13 minutes)

Students work independently for 8–10 minutes. Allow 3–5 minutes to go over the answers as a whole group.

Exercises 9–19

9. Meg is playing a card game with her friend, Iona. The cards have positive and negative numbers printed on them. Meg exclaims: "The absolute value of the number on my card equals 8." What is the number on Meg's card?

    $|-8| = 8$ or $|8| = 8$

    Meg either has 8 or −8 on her card.

10. List a positive and negative number whose absolute value is greater than 3. Justify your answer using the number line.

    Answers may vary. $|-4| = 4$ and $|7| = 7$; $4 > 3$ and $7 > 3$. On a number line, the distance from zero to -4 is 4 units. So, the absolute value of -4 is 4. The number 4 is to the right of 3 on the number line, so 4 is greater than 3. The distance from zero to 7 on a number line is 7 units, so the absolute value of 7 is 7. Since 7 is to the right of 3 on the number line, 7 is greater than 3.

11. Which of the following situations can be represented by the absolute value of 10? Check all that apply.

    _____ The temperature is 10 degrees below zero. Express this as an integer.

    __X__ Determine the size of Harold's debt if he owes $10.

    __X__ Determine how far -10 is from zero on a number line.

    __X__ 10 degrees is how many degrees above zero?

12. Julia used absolute value to find the distance between 0 and 6 on a number line. She then wrote a similar statement to represent the distance between 0 and −6. Below is her work. Is it correct? Explain.

    $$|6| = 6 \text{ and } |-6| = -6$$

    No. The distance is 6 units whether you go from 0 to 6 or 0 to -6. So, the absolute value of -6 should also be 6, but Julia said it was -6.

13. Use absolute value to represent the amount, in dollars, of a $238.25 profit.

    $|238.25| = 238.25$

14. Judy lost 15 pounds. Use absolute value to represent the number of pounds Judy lost.

    $|-15| = 15$

15. In math class, Carl and Angela are debating about integers and absolute value. Carl said two integers can have the same absolute value, and Angela said one integer can have two absolute values. Who is right? Defend your answer.

    Carl is right. An integer and its opposite are the same distance from zero. So, they have the same absolute values because absolute value is the distance between the number and zero.

Lesson 11: Absolute Value—Magnitude and Distance

> 16. Jamie told his math teacher: "Give me any absolute value, and I can tell you two numbers that have that absolute value." Is Jamie correct? For any given absolute value, will there always be two numbers that have that absolute value?
>
>   *No, Jamie is not correct because zero is its own opposite. Only one number has an absolute value of 0, and that would be 0.*
>
> 17. Use a number line to show why a number and its opposite have the same absolute value.
>
>   *A number and its opposite are the same distance from zero but on opposite sides. An example is 5 and −5. These numbers are both 5 units from zero. Their distance is the same, so they have the same absolute value, 5.*
>
>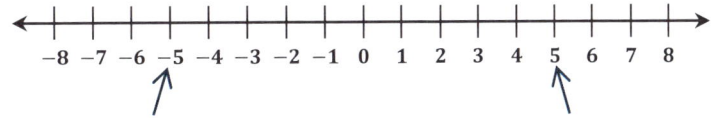
>
> 18. A bank teller assisted two customers with transactions. One customer made a $25 withdrawal from a savings account. The other customer made a $15 deposit. Use absolute value to show the size of each transaction. Which transaction involved more money?
>
> $|-25| = 25$ *and* $|15| = 15$. *The $25 withdrawal involved more money.*
>
> 19. Which is farther from zero: $-7\frac{3}{4}$ or $7\frac{1}{2}$? Use absolute value to defend your answer.
>
>   *The number that is farther from 0 is* $-7\frac{3}{4}$. *This is because* $\left|-7\frac{3}{4}\right| = 7\frac{3}{4}$ *and* $\left|7\frac{1}{2}\right| = 7\frac{1}{2}$. *Absolute value is a number's distance from zero. I compared the absolute value of each number to determine which was farther from zero. The absolute value of* $-7\frac{3}{4}$ *is* $7\frac{3}{4}$. *The absolute value of* $7\frac{1}{2}$ *is* $7\frac{1}{2}$. *We know that* $7\frac{3}{4}$ *is greater than* $7\frac{1}{2}$. *Therefore,* $-7\frac{3}{4}$ *is farther from zero than* $7\frac{1}{2}$.

Closing (3 minutes)

- I am thinking of two numbers. Both numbers have the same absolute value. What must be true about the two numbers?
    - *The numbers are opposites.*
- Can the absolute value of a number ever be a negative number? Why or why not?
    - *No. Absolute value is the distance a number is from zero. If you count the number of units from zero to the number, the number of units is its absolute value. You could be on the right or left side of zero, but the number of units you count represents the distance or absolute value, and that will always be a positive number.*
- How can we use absolute value to determine magnitude? For instance, how far below zero is −8 degrees?
    - *Absolute value represents magnitude. This means that −8 degrees is 8 units below zero.*

Exit Ticket (6 minutes)

Lesson 11: Absolute Value—Magnitude and Distance

Exit Ticket

Jessie and his family drove up to a picnic area on a mountain. In the morning, they followed a trail that led to the mountain summit, which was 2,000 feet above the picnic area. They then returned to the picnic area for lunch. After lunch, they hiked on a trail that led to the mountain overlook, which was 3,500 feet below the picnic area.

a. Locate and label the elevation of the mountain summit and mountain overlook on a vertical number line. The picnic area represents zero. Write a rational number to represent each location.

   Picnic area: 0

   Mountain summit: _____

   Mountain overlook: _____

b. Use absolute value to represent the distance on the number line of each location from the picnic area.

   Distance from the picnic area to the mountain summit: _____

   Distance from the picnic area to the mountain overlook: _____

c. What is the distance between the elevations of the summit and overlook? Use absolute value and your number line from part (a) to explain your answer.

A STORY OF RATIOS   Lesson 11   6•3

Exit Ticket Sample Solutions

Jessie and his family drove up to a picnic area on a mountain. In the morning, they followed a trail that led to the mountain summit, which was 2,000 feet above the picnic area. They then returned to the picnic area for lunch. After lunch, they hiked on a trail that led to the mountain overlook, which was 3,500 feet below the picnic area.

a. Locate and label the elevation of the mountain summit and mountain overlook on a vertical number line. The picnic area represents zero. Write a rational number to represent each location.

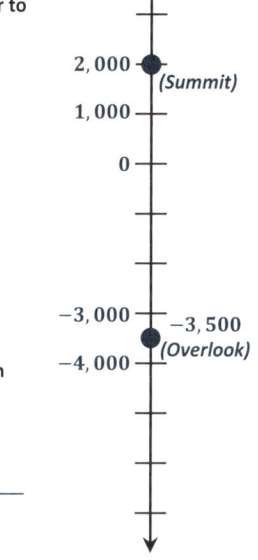

Picnic area: __0__

Mountain summit: __2,000__

Mountain overlook: __−3,500__

b. Use absolute value to represent the distance on the number line of each location from the picnic area.

Distance from the picnic area to the mountain summit: __|2,000| = 2,000__

Distance from the picnic area to the mountain overlook: __|−3,500| = 3,500__

c. What is the distance between the elevations of the summit and overlook? Use absolute value and your number line from part (a) to explain your answer.

Summit to picnic area and picnic area to overlook: $2,000 + 3,500 = 5,500$

There are $2,000$ units from zero to $2,000$ on the number line.

There are $3,500$ units from zero to $-3,500$ on the number line.

Altogether, that equals $5,500$ units, which represents the distance on the number line between the two elevations. Therefore, the difference in elevations is $5,500$ feet.

Problem Set Sample Solutions

For each of the following two quantities in Problems 1–4, which has the greater magnitude? (Use absolute value to defend your answers.)

1. 33 dollars and −52 dollars

   $|-52| = 52$ $|33| = 33$ *$52 > 33$, so -52 dollars has the greater magnitude.*

2. −14 feet and 23 feet

   $|-14| = 14$ $|23| = 23$ *$14 < 23$, so 23 feet has the greater magnitude.*

Lesson 11:   Absolute Value—Magnitude and Distance   107

This work is derived from Eureka Math ™ and licensed by Great Minds. ©2015 Great Minds. eureka-math.org
G6-M3-TE-B3-1.3.0-07.2015

3. -24.6 pounds and -24.58 pounds

   $|-24.6| = 24.6$ $|-24.58| = 24.58$ $24.6 > 24.58$, so -24.6 pounds has the greater magnitude.

4. $-11\frac{1}{4}$ degrees and 11 degrees

   $\left|-11\frac{1}{4}\right| = 11\frac{1}{4}$ $|11| = 11$ $11\frac{1}{4} > 11$, so $-11\frac{1}{4}$ degrees has the greater magnitude.

For Problems 5–7, answer true or false. If false, explain why.

5. The absolute value of a negative number will always be a positive number.

   True

6. The absolute value of any number will always be a positive number.

   False. Zero is the exception since the absolute value of zero is zero, and zero is not positive.

7. Positive numbers will always have a higher absolute value than negative numbers.

   False. A number and its opposite have the same absolute value.

8. Write a word problem whose solution is $|20| = 20$.

   Answers will vary. Kelli flew a kite 20 feet above the ground. Determine the distance between the kite and the ground.

9. Write a word problem whose solution is $|-70| = 70$.

   Answers will vary. Paul dug a hole in his yard 70 inches deep to prepare for an in-ground swimming pool. Determine the distance between the ground and the bottom of the hole that Paul dug.

10. Look at the bank account transactions listed below, and determine which has the greatest impact on the account balance. Explain.

    a. A withdrawal of $\$60$
    b. A deposit of $\$55$
    c. A withdrawal of $\$58.50$

    $|-60| = 60$ $|55| = 55$ $|-58.50| = 58.50$

    $60 > 58.50 > 55$, so a withdrawal of $\$60$ has the greatest impact on the account balance.

Lesson 11: Absolute Value—Magnitude and Distance

A STORY OF RATIOS   Lesson 12   6•3

Lesson 12: The Relationship Between Absolute Value and Order

Student Outcomes

- Students understand that the order of positive numbers is the same as the order of their absolute values.
- Students understand that the order of negative numbers is the opposite order of their absolute values.
- Students understand that negative numbers are always less than positive numbers.

Lesson Notes

Prior to presenting the lesson, prepare sticky notes containing a balanced variety of positive and negative integers ranging from -50 to 50. Each pair of students requires a set of ten integers including five negative values, zero, and four positive values.

Classwork

Opening Exercise (5 minutes)

Divide students into pairs. Provide each pair with a set of ten integers including five negative values, zero, and four positive values written on sticky notes. Instruct student groups to order their integers from least to greatest based on prior knowledge. When the integers are in the correct order, students record the integer values in order in their student materials.

> **Opening Exercise**
>
> Record your integer values in order from least to greatest in the space below.
>
> Sample answer: $-12, -9, -5, -2, -1, 0, 2, 5, 7, 8$

Have one pair of students post their sticky notes to the wall in the specified order. Ask another pair of students:

- Explain the reasoning for the order.
    - *The integers are in the same order in which they would be found located from left (or bottom) to right (or top) on the number line.*

A STORY OF RATIOS   Lesson 12 6•3

Example 1 (8 minutes): Comparing Order of Integers to the Order of Their Absolute Values

Students use the integer values from the Opening Exercise to compare the order of integers to the order of their absolute values.

> **Example 1: Comparing Order of Integers to the Order of Their Absolute Values**
>
> Write an inequality statement relating the ordered integers from the Opening Exercise. Below each integer, write its absolute value.
>
> Sample answer: $-12 < -9 < -5 < -2 < -1 < 0 < 2 < 5 < 7 < 8$
>
> 12 $\phantom{<}$ 9 $\phantom{<}$ 5 $\phantom{<}$ 2 $\phantom{<}$ 1 $\phantom{<}$ 0 $\phantom{<}$ 2 $\phantom{<}$ 5 $\phantom{<}$ 7 $\phantom{<}$ 8

Scaffolding:

For English Language Learners: In place of the words *ascending* and *descending*, describe numbers as increasing from left to right or decreasing from left to right.

- Are the absolute values of your integers in order? Explain.
  - No. The absolute values of the positive integers listed to the right of zero are still in ascending order, but the absolute values of the negative integers listed to the left of zero are now in descending order.

MP.7

> Circle the absolute values that are in increasing numerical order and their corresponding integers. Describe the circled values.
>
> $-12 < -9 < -5 < -2 < -1 < 0 < 2 < 5 < 7 < 8$
>
> 12 $\phantom{<}$ 9 $\phantom{<}$ 5 $\phantom{<}$ 2 $\phantom{<}$ 1 $\phantom{<}$ 0 $\phantom{<}$ 2 $\phantom{<}$ 5 $\phantom{<}$ 7 $\phantom{<}$ 8
>
> *The circled integers are all positive values except zero. The positive integers and their absolute values have the same order.*
>
> Rewrite the integers that are not circled in the space below. How do these integers differ from the ones you circled?
>
> $-12, -9, -5, -2, -1$
>
> *They are all negative integers.*
>
> Rewrite the negative integers in ascending order and their absolute values in ascending order below them.
>
> $-12 < -9 < -5 < -2 < -1$
>
> 1 $\phantom{<}$ 2 $\phantom{<}$ 5 $\phantom{<}$ 9 $\phantom{<}$ 12
>
> Describe how the order of the absolute values compares to the order of the negative integers.
>
> *The orders of the negative integers and their corresponding absolute values are opposite.*

Example 2 (8 minutes): The Order of Negative Integers and Their Absolute Values

Students examine the lengths of arrows corresponding with positive and negative integers on the number line and use their analysis to understand why the order of negative integers is opposite the order of their absolute values. Monitor the room, and provide guidance as needed, and then guide a whole-class discussion with questions.

Lesson 12: The Relationship Between Absolute Value and Order

A STORY OF RATIOS   Lesson 12   6•3

> **Example 2: The Order of Negative Integers and Their Absolute Values**
>
> Draw arrows starting at the dashed line (zero) to represent each of the integers shown on the number line below. The arrows that correspond with 1 and 2 have been modeled for you.
>
>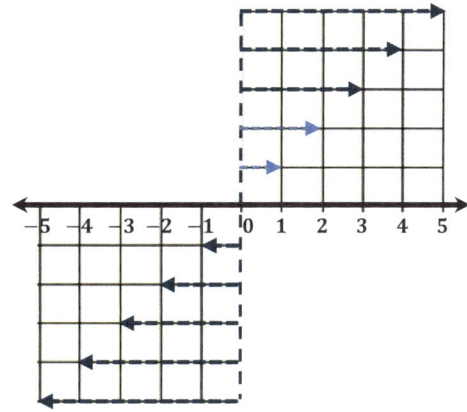

MP.2

- Which integer(s) is represented by the longest arrow that you drew? Why?
    - *5 and −5 because, of the integers shown, they are farthest from zero on the number line.*
- Which nonzero integer(s) is represented by the shortest arrow? Why?
    - *1 and −1 because, of the integers shown, they are closest to zero on the number line.*
- Is there an arrow for the integer 0? Explain.
    - *The length of such an arrow would be 0, so we could not see the arrow. We could call it an arrow with zero length, but we could not draw it.*
- Do these "arrows" start at the same place on the number line? Why or why not?
    - *They all start at zero on the number line because 0 is the reference point for all numbers on the number line.*
- Do all arrows point in the same direction? Why or why not?
    - *All arrows do not point in the same direction because some integers have opposite signs. The lengths of the arrows get shorter as you approach zero from the left or from the right, which means the absolute values decrease as you approach zero from the left or the right.*

Direct students to complete the statements in their student materials by filling in the blanks.

> As you approach zero from the left on the number line, the integers ___increase___, but the absolute values of those integers ___decrease___. This means that the order of negative integers is ___opposite___ the order of their absolute values.

Lesson 12: The Relationship Between Absolute Value and Order   111

A STORY OF RATIOS   Lesson 12   6•3

Discussion (2 minutes)

- Think of a real-world example that shows why the order of negative integers and the order of their absolute values are opposite.
  - *Alec, Benny, and Charlotte have cafeteria charge account balances of -3, -4, and -5 dollars, respectively. The number that represents Alec's cafeteria charge account balance is the greatest because he is the least in debt. Alec owes the least amount of money and is the closest to having a positive balance. His balance of -3 dollars is farthest right on the number line of the three balances.*

Exercise 1 (5 minutes)

Students separate positive and negative rational numbers and order them according to their absolute values. Students then order the given set of positive and negative rational numbers using the orders of their absolute values.

Exercise 1

Complete the steps below to order these numbers:

$$\left\{2.1, -4\frac{1}{2}, -6, 0.25, -1.5, 0, 3.9, -6.3, -4, 2\frac{3}{4}, 3.99, -9\frac{1}{4}\right\}$$

a. Separate the set of numbers into positive rational numbers, negative rational numbers, and zero in the top cells below (order does not matter).

b. Write the absolute values of the rational numbers (order does not matter) in the bottom cells below.

Negative Rational Numbers

$-4\frac{1}{2}$ -6 -6.3

-4 $-9\frac{1}{4}$ -1.5

Zero

0

Positive Rational Numbers

2.1 0.25 3.9

$2\frac{3}{4}$ 3.99

Absolute Values

$4\frac{1}{2}$ 6 6.3

4 $9\frac{1}{4}$ 1.5

Absolute Values

2.1 0.25 3.9

$2\frac{3}{4}$ 3.99

c. Order each subset of absolute values from least to greatest.

$1.5, \ 4, \ 4\frac{1}{2}, \ 6, \ 6.3, \ 9\frac{1}{4}$ 0 $0.25, \ 2.1, \ 2\frac{3}{4}, \ 3.9, \ 3.99$

d. Order each subset of rational numbers from least to greatest.

$-9\frac{1}{4}, \ -6.3, \ -6, \ -4\frac{1}{2}, \ -4, \ -1.5$ 0 $0.25, \ 2.1, \ 2\frac{3}{4}, \ 3.9, \ 3.99$

MP.2

A STORY OF RATIOS Lesson 12 6•3

e. Order the whole given set of rational numbers from least to greatest.

$$-9\tfrac{1}{4},\ -6.3,\ -6,\ -4\tfrac{1}{2},\ -4,\ -1.5,\ 0,\ 0.25,\ 2.1,\ 2\tfrac{3}{4},\ 3.9,\ 3.99$$

Exercise 2 (8 minutes)

Students independently complete the following problems. Monitor student progress, and provide guidance as needed.

Exercise 2

a. Find a set of four integers such that their order and the order of their absolute values are the same.

Answers will vary. An example follows: 4, 6, 8, 10

b. Find a set of four integers such that their order and the order of their absolute values are opposite.

Answers will vary. An example follows: −10, −8, −6, −4

c. Find a set of four non-integer rational numbers such that their order and the order of their absolute values are the same.

Answers will vary. An example follows: $2\tfrac{1}{2},\ 3\tfrac{1}{2},\ 4\tfrac{1}{2},\ 5\tfrac{1}{2}$

d. Find a set of four non-integer rational numbers such that their order and the order of their absolute values are opposite.

Answers will vary. An example follows: $-5\tfrac{1}{2},\ -4\tfrac{1}{2},\ -3\tfrac{1}{2},\ -2\tfrac{1}{2}$

e. Order all of your numbers from parts (a)–(d) in the space below. This means you should be ordering 16 numbers from least to greatest.

Answers will vary. An example follows:

$$-10,\ -8,\ -6,\ -5\tfrac{1}{2},\ -4\tfrac{1}{2},\ -4,\ -3\tfrac{1}{2},\ -2\tfrac{1}{2},\ 2\tfrac{1}{2},\ 3\tfrac{1}{2},\ 4,\ 4\tfrac{1}{2},\ 5\tfrac{1}{2},\ 6,\ 8,\ 10$$

Closing (4 minutes)

- Below are the absolute values of a set of rational numbers in increasing order.

$\{0.4,\ 1,\ 2\tfrac{1}{2},\ 4,\ 4.3,\ 7,\ 9.9\}$

- Can you determine the order of the rational numbers themselves? If so, explain why, and write the numbers in increasing order. If not, explain why.
  - *It is not possible to determine the order of the rational numbers because we do not know the signs of the rational numbers.*

Lesson 12: The Relationship Between Absolute Value and Order

113

This work is derived from Eureka Math ™ and licensed by Great Minds. ©2015 Great Minds. eureka-math.org
G6-M3-TE-B3-1.3.0-07.2015

- If you are told that the numbers above represent the absolute values of positive rational numbers, can you determine the order of the rational numbers? If so, explain why, and write the numbers in increasing order. If not, explain why not.
  - *If the original numbers are all positive, we are able to order the rational numbers because we know their signs. The order of the original numbers will be the same as the order of their absolute values.*

- If you are told that the numbers above represent the absolute values of negative rational numbers, can you find the order of the rational numbers? If so, explain why, and write the numbers in increasing order. If not, explain why.
  - *If the original numbers are all negative, we are able to order the rational numbers because we know their signs. The order of the original numbers will be the opposite order of their absolute values.*

> **Lesson Summary**
>
> The absolute values of positive numbers always have the same order as the positive numbers themselves. Negative numbers, however, have exactly the opposite order as their absolute values. The absolute values of numbers on the number line increase as you move away from zero in either direction.

Exit Ticket (5 minutes)

A STORY OF RATIOS   Lesson 12   6•3

Name _____   Date _____

Lesson 12: The Relationship Between Absolute Value and Order

Exit Ticket

1. Bethany writes a set of rational numbers in increasing order. Her teacher asks her to write the absolute values of these numbers in increasing order. When her teacher checks Bethany's work, she is pleased to see that Bethany has not changed the order of her numbers. Why is this?

2. Mason was ordering the following rational numbers in math class: $-3.3, \ -15, \ -8\frac{8}{9}$.

   a. Order the numbers from least to greatest.

   b. List the order of their absolute values from least to greatest.

   c. Explain why the orderings in parts (a) and (b) are different.

Lesson 12:   The Relationship Between Absolute Value and Order   115

A STORY OF RATIOS — Lesson 12 — 6•3

Exit Ticket Sample Solutions

1. Bethany writes a set of rational numbers in increasing order. Her teacher asks her to write the absolute values of these numbers in increasing order. When her teacher checks Bethany's work, she is pleased to see that Bethany has not changed the order of her numbers. Why is this?

   All of Bethany's rational numbers are positive or 0. The positive rational numbers have the same order as their absolute values. If any of Bethany's rational numbers are negative, then the order would be different.

2. Mason was ordering the following rational numbers in math class: $-3.3,\ -15,\ -8\frac{8}{9}$.

   a. Order the numbers from least to greatest.

   $-15,\ -8\frac{8}{9},\ -3.3$

   b. List the order of their absolute values from least to greatest.

   $3.3,\ 8\frac{8}{9},\ 15$

   c. Explain why the orderings in parts (a) and (b) are different.

   Since these are all negative numbers, when I ordered them from least to greatest, the one farthest away from zero (farthest to the left on the number line) came first. This number is -15. Absolute value is the numbers' distance from zero, and so the number farthest away from zero has the greatest absolute value, so 15 will be greatest in the list of absolute values, and so on.

Problem Set Sample Solutions

1. Micah and Joel each have a set of five rational numbers. Although their sets are not the same, their sets of numbers have absolute values that are the same. Show an example of what Micah and Joel could have for numbers. Give the sets in order and the absolute values in order.

   Examples may vary. If Micah had $1,\ 2,\ 3,\ 4,\ 5$, then his order of absolute values would be the same: $1,\ 2,\ 3,\ 4,\ 5$. If Joel had the numbers $-5,\ -4,\ -3,\ -2,\ -1$, then his order of absolute values would also be $1,\ 2,\ 3,\ 4,\ 5$.

   Enrichment Extension: Show an example where Micah and Joel both have positive and negative numbers.

   If Micah had the numbers: $-5,\ -3,\ -1,\ 2,\ 4$, his order of absolute values would be $1,\ 2,\ 3,\ 4,\ 5$. If Joel had the numbers $-4,\ -2,\ 1,\ 3,\ 5$, then the order of his absolute values would also be $1,\ 2,\ 3,\ 4,\ 5$.

Lesson 12: The Relationship Between Absolute Value and Order

2. For each pair of rational numbers below, place each number in the Venn diagram based on how it compares to the other.

   a. $-4, -8$
   b. $4, 8$
   c. $7, -3$
   d. $-9, 2$
   e. $6, 1$
   f. $-5, 5$
   g. $-2, 0$

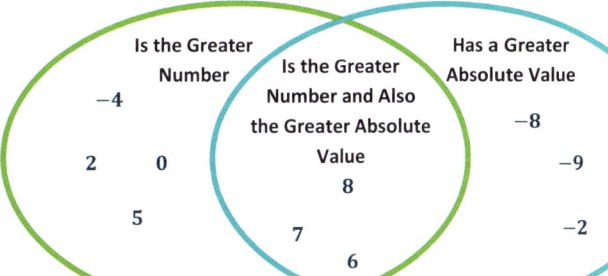

Lesson 13: Statements of Order in the Real World

Student Outcomes

- Students apply understanding of order and absolute value when examining real-world scenarios. Students realize, for instance, that the depth of a location below sea level is the absolute value of a negative number, while the height of an object above sea level is the absolute value of a positive number.

Classwork

Opening Exercise (4 minutes)

Students discuss the meaning of the report below, write a summary of their conclusions in their student materials, and provide their feedback to the whole group.

> **Opening Exercise**
>
> A radio disc jockey reports that the temperature outside his studio has changed 10 degrees since he came on the air this morning. Discuss with your group what listeners can conclude from this report.
>
> *The report is not specific enough to be conclusive because 10 degrees of change could mean an increase or a decrease in temperature. A listener might assume the report says an increase in temperature; however, the word "changed" is not specific enough to conclude a positive or negative change.*

- How could you change the report to make it more informative?
    - *Using the words "increased" or "decreased" instead of "changed" would be much more informative.*
- In real-world contexts, descriptive words such as *debt, credit, increase,* and *decrease* help us indicate when a given magnitude is representative of a positive or negative value.

Example 1 (4 minutes): Ordering Numbers in the Real World

Students saw that absolute value represents the magnitude of a positive or negative quantity in Lesson 11. To order rational numbers given in a real-world context, students need to consider the meaning of descriptors and what they indicate about a given quantity.

Scaffolding:
Review the financial terms if necessary. Have students develop a poster to categorize the terms and include examples of their meanings.

> **Example 1: Ordering Numbers in the Real World**
>
> A $25 credit and a $25 charge appear similar, yet they are very different.
>
> Describe what is similar about the two transactions.
>
> *The transactions look similar because they are described using the same number. Both transactions have the same magnitude (or absolute value) and, therefore, result in a change of $25 to an account balance.*

> How do the two transactions differ?
>
> *The credit would cause an increase to an account balance and, therefore, should be represented by 25, while the charge would instead decrease an account balance and should be represented by −25. The two transactions represent changes that are opposites.*

Exercises 1–4 (22 minutes)

Students use absolute value to solve various real-world problems with partners and then share and support their conclusions with the whole group.

Allow two minutes for setup. Have eight students arrange their desks into two rows of four so that the rows are facing each other. Additional groups of eight students should be formed per individual class size. If using tables, have four students sit on one side of the table(s) and the other four students sit opposite them. Students rotate when each problem is completed. When the rotation occurs, students who are in the right-most seat of each row rotate to a position in the opposite row (see diagram to the right). Having students move in opposite directions allows each student to work with a different partner on each of the four problems.

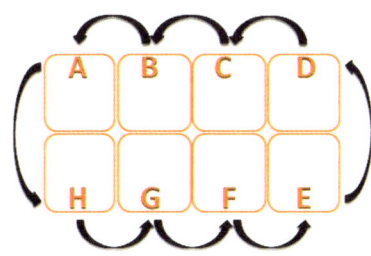

Students should work at their stations for 15 minutes, completing Exercise 1 in the student materials with their first partner, Exercise 2 with a new partner, Exercise 3 with a different new partner, and finally Exercise 4 with a different new partner. Partners are given three minutes to complete each problem in the student materials and then given one minute to rotate seats.

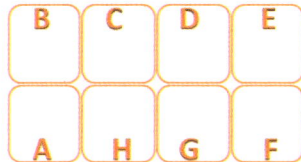

Each problem requires that students determine appropriate rational numbers to represent given quantities and order the numbers as specified. Students also provide reasoning for their choices of rational numbers in each case.

Exercises

1. Scientists are studying temperatures and weather patterns in the Northern Hemisphere. They recorded temperatures (in degrees Celsius) in the table below as reported in emails from various participants. Represent each reported temperature using a rational number. Order the rational numbers from least to greatest. Explain why the rational numbers that you chose appropriately represent the given temperatures.

Temperatures as Reported	8 below zero	12	−4	13 below zero	0	2 above zero	6 below zero	−5
Temperature (°C)	−8	12	−4	−13	0	2	−6	−5

$-13 < -8 < -6 < -5 < -4 < 0 < 2 < 12$

The words "below zero" refer to negative numbers because they are located below zero on a vertical number line.

Lesson 13: Statements of Order in the Real World

A STORY OF RATIOS   Lesson 13   6•3

2. Jami's bank account statement shows the transactions below. Represent each transaction as a rational number describing how it changes Jami's account balance. Then, order the rational numbers from greatest to least. Explain why the rational numbers that you chose appropriately reflect the given transactions.

Listed Transactions	Debit $12.20	Credit $4.08	Charge $1.50	Withdrawal $20.00	Deposit $5.50	Debit $3.95	Charge $3.00
Change to Jami's Account	−12.2	4.08	−1.5	−20	5.5	−3.95	−3

$5.5 > 4.08 > -1.5 > -3 > -3.95 > -12.2 > -20$

The words "debit," "charge," and "withdrawal" all describe transactions in which money is taken out of Jami's account, decreasing its balance. These transactions are represented by negative numbers. The words "credit" and "deposit" describe transactions that will put money into Jami's account, increasing its balance. These transactions are represented by positive numbers.

3. During the summer, Madison monitors the water level in her parents' swimming pool to make sure it is not too far above or below normal. The table below shows the numbers she recorded in July and August to represent how the water levels compare to normal. Order the rational numbers from least to greatest. Explain why the rational numbers that you chose appropriately reflect the given water levels.

Madison's Readings	$\frac{1}{2}$ inch above normal	$\frac{1}{4}$ inch above normal	$\frac{1}{2}$ inch below normal	$\frac{1}{8}$ inch above normal	$1\frac{1}{4}$ inches below normal	$\frac{3}{8}$ inch below normal	$\frac{3}{4}$ inch below normal
Compared to Normal	$\frac{1}{2}$	$\frac{1}{4}$	$-\frac{1}{2}$	$\frac{1}{8}$	$-1\frac{1}{4}$	$-\frac{3}{8}$	$-\frac{3}{4}$

$-1\frac{1}{4} < -\frac{3}{4} < -\frac{1}{2} < -\frac{3}{8} < \frac{1}{8} < \frac{1}{4} < \frac{1}{2}$

The measurements are taken in reference to normal level, which is considered to be 0. The words "above normal" refer to the positive numbers located above zero on a vertical number line, and the words "below normal" refer to the negative numbers located below zero on a vertical number line.

4. Changes in the weather can be predicted by changes in the barometric pressure. Over several weeks, Stephanie recorded changes in barometric pressure seen on her barometer to compare to local weather forecasts. Her observations are recorded in the table below. Use rational numbers to record the indicated changes in the pressure in the second row of the table. Order the rational numbers from least to greatest. Explain why the rational numbers that you chose appropriately represent the given pressure changes.

Barometric Pressure Change (Inches of Mercury)	Rise 0.04	Fall 0.21	Rise 0.2	Fall 0.03	Rise 0.1	Fall 0.09	Fall 0.14
Barometric Pressure Change (Inches of Mercury)	0.04	−0.21	0.2	−0.03	0.1	−0.09	−0.14

$-0.21, -0.14, -0.09, -0.03, 0.04, 0.1, 0.2$

The records that include the word "rise" refer to increases and are, therefore, represented by positive numbers. The records that include the word "fall" refer to decreases and are, therefore, represented by negative numbers.

After completing all stations, ask students to report their answers and reasoning for the given problems. Look for differences in valid reasoning, and discuss those differences where appropriate. Encourage students to politely challenge the reasoning of their classmates if applicable. This activity should take four minutes.

A STORY OF RATIOS   Lesson 13   6•3

Example 2 (5 minutes): Using Absolute Value to Solve Real-World Problems

Students use the absolute values of positive and negative numbers to solve problems in real-world contexts. Students may find it helpful to draw a picture as a problem-solving strategy. Grid paper may be provided so that they can accurately construct a picture and number line diagram.

> **Example 2: Using Absolute Value to Solve Real-World Problems**
>
> The captain of a fishing vessel is standing on the deck at 23 feet above sea level. He holds a rope tied to his fishing net that is below him underwater at a depth of 38 feet.
>
> Draw a diagram using a number line, and then use absolute value to compare the lengths of rope in and out of the water.
>
> *The captain is above the water, and the fishing net is below the water's surface. Using the water level as reference point zero, I can draw the diagram using a vertical number line. The captain is located at 23, and the fishing net is located at −38.*
>
> $|23| = 23$ and $|-38| = 38$, so there is more rope underwater than above.
>
> $38 - 23 = 15$
>
> *The length of rope below the water's surface is 15 feet longer than the rope above water.*

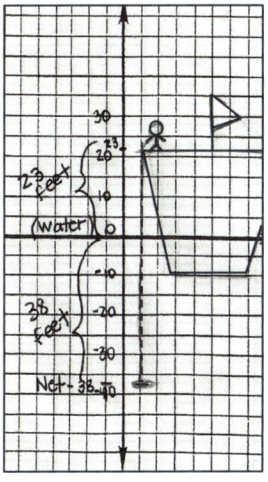

Discuss and model how students can construct a number line and use number sense to find the approximate location of 23 and −38 without grid paper.

Example 3 (4 minutes): Making Sense of Absolute Value and Statements of Inequality

Students examine absolute values of negative numbers in a real-world context and make sense of statements about inequalities involving those values.

> **Example 3: Making Sense of Absolute Value and Statements of Inequality**
>
> A recent television commercial asked viewers, "Do you have over $10,000 in credit card debt?"
>
> What types of numbers are associated with the word *debt*, and why? Write a number that represents the value from the television commercial.
>
> *Negative numbers; debt describes money that is owed; −10,000*
>
> Give one example of "over $10,000 in credit card debt." Then, write a rational number that represents your example.
>
> *Answers will vary, but the number should have a value of less than −10,000. Credit card debt of $11,000; −11,000*

Lesson 13:   Statements of Order in the Real World   121

> How do the debts compare, and how do the rational numbers that describe them compare? Explain.
>
> *The example $11,000 is greater than $10,000 from the commercial; however, the rational numbers that represent these debt values have the opposite order because they are negative numbers. $-11,000 < -10,000$. The absolute values of negative numbers have the opposite order of the negative values themselves.*

Closing (3 minutes)

- Your friend Samuel says he is 50 feet from sea level. What additional information should Samuel give you in order for you to identify his elevation?
    - *In order to know Samuel's elevation, he would have to tell me if he is above or below sea level.*
- Identify three real-world situations that are represented by negative rational numbers.
    - *The temperature is 3°F below zero. My mom was charged a $15 fee for missing a doctor appointment. Jason went scuba diving and was 20 feet below sea level.*
- Identify three real-world situations that are represented by positive rational numbers.
    - *The temperature is 3°F above zero. My mom received a $15 credit for referring a friend to her Internet service. Jason went hiking and was 20 feet above sea level.*

> **Lesson Summary**
>
> When comparing values in real-world situations, descriptive words help you to determine if the number represents a positive or negative number. Making this distinction is critical when solving problems in the real world. Also critical is to understand how an inequality statement about an absolute value compares to an inequality statement about the number itself.

Exit Ticket (3 minutes)

Name _____ Date _____

Lesson 13: Statements of Order in the Real World

Exit Ticket

1. Loni and Daryl call each other from different sides of Watertown. Their locations are shown on the number line below using miles. Use absolute value to explain who is a farther distance (in miles) from Watertown. How much closer is one than the other?

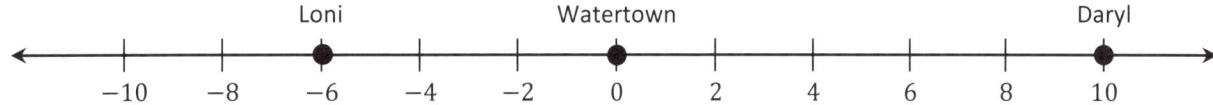

2. Claude recently read that no one has ever scuba dived more than 330 meters below sea level. Describe what this means in terms of elevation using sea level as a reference point.

Exit Ticket Sample Solutions

1. Loni and Daryl call each other from different sides of Watertown. Their locations are shown on the number line below using miles. Use absolute value to explain who is a farther distance (in miles) from Watertown. How much closer is one than the other?

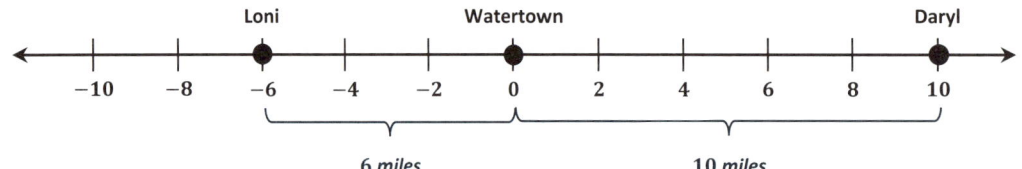

   Loni's location is -6, and $|-6| = 6$ because -6 is 6 units from 0 on the number line. Daryl's location is 10, and $|10| = 10$ because 10 is 10 units from 0 on the number line. We know that $10 > 6$, so Daryl is farther from Watertown than Loni.

   $10 - 6 = 4$; Loni is 4 miles closer to Watertown than Daryl.

2. Claude recently read that no one has ever scuba dived more than 330 meters below sea level. Describe what this means in terms of elevation using sea level as a reference point.

   330 meters below sea level is an elevation of -330 feet. "More than 330 meters below sea level" means that no diver has ever had more than 330 meters between himself and sea level when he was below the water's surface while scuba diving.

Problem Set Sample Solutions

1. Negative air pressure created by an air pump makes a vacuum cleaner able to collect air and dirt into a bag or other container. Below are several readings from a pressure gauge. Write rational numbers to represent each of the readings, and then order the rational numbers from least to greatest.

Gauge Readings (pounds per square inch)	25 psi pressure	13 psi vacuum	6.3 psi vacuum	7.8 psi vacuum	1.9 psi vacuum	2 psi pressure	7.8 psi pressure
Pressure Readings (pounds per square inch)	25	-13	-6.3	-7.8	-1.9	2	7.8

$$-13 < -7.8 < -6.3 < -1.9 < 2 < 7.8 < 25$$

2. The fuel gauge in Nic's car says that he has 26 miles to go until his tank is empty. He passed a fuel station 19 miles ago, and a sign says there is a town only 8 miles ahead. If he takes a chance and drives ahead to the town and there isn't a fuel station there, does he have enough fuel to go back to the last station? Include a diagram along a number line, and use absolute value to find your answer.

No, he does not have enough fuel to drive to the town and then back to the fuel station. He needs 8 miles' worth of fuel to get to the town, which lowers his limit to 18 miles. The total distance between the fuel station and the town is 27 miles; $|8| + |-19| = 8 + 19 = 27$. Nic would be 9 miles short on fuel. It would be safer to go back to the fuel station without going to the town first.

Lesson 13: Statements of Order in the Real World

Name _____ Date _____

1. The picture below is a flood gauge that is used to measure how far (in feet) a river's water level is above or below its normal level.

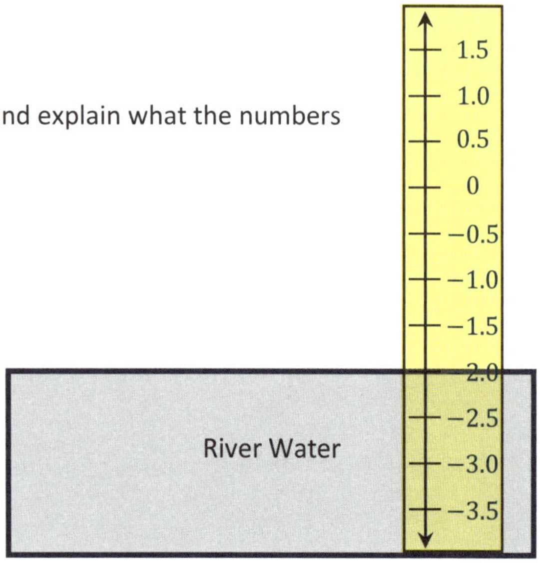

   a. Explain what the number 0 on the gauge represents, and explain what the numbers above and below 0 represent.

   b. Describe what the picture indicates about the river's current water level.

   c. What number represents the opposite of the water level shown in the picture, and where is it located on the gauge? What would it mean if the river water was at that level?

   d. If heavy rain is in the forecast for the area for the next 24 hours, what reading might you expect to see on this gauge tomorrow? Explain your reasoning.

2. Isaac made a mistake in his checkbook. He wrote a check for $8.98 to rent a video game but mistakenly recorded it in his checkbook as an $8.98 deposit.

   a. Represent each transaction with a rational number, and explain the difference between the transactions.

   b. On the number line below, locate and label the points that represent the rational numbers listed in part (a). Describe the relationship between these two numbers. Zero on the number line represents Isaac's balance before the mistake was made.

   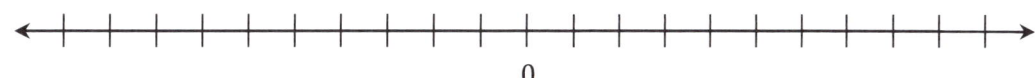

   c. Use absolute value to explain how a debit of $8.98 and a credit of $8.98 are similar.

A STORY OF RATIOS   Mid-Module Assessment Task   6•3

3. A local park's programs committee is raising money by holding mountain bike races on a course through the park. During each race, a computer tracks the competitors' locations on the course using GPS tracking. The table shows how far each competitor is from a checkpoint.

Number	Competitor Name	Distance to Checkpoint
223	Florence	0.1 miles before
231	Mary	$\frac{2}{5}$ miles past
240	Rebecca	0.5 miles before
249	Lita	$\frac{1}{2}$ miles past
255	Nancy	$\frac{2}{10}$ miles before

a. The checkpoint is represented by 0 on the number line. Locate and label points on the number line for the positions of each listed participant. Label the points using rational numbers.

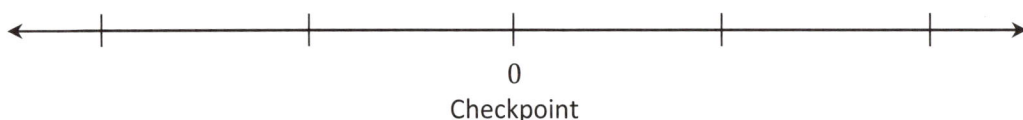
0
Checkpoint

b. Which of the competitors is closest to the checkpoint? Explain.

c. Two competitors are the same distance from the checkpoint. Are they in the same location? Explain.

d. Who is closer to finishing the race, Nancy or Florence? Support your answer.

A STORY OF RATIOS   Mid-Module Assessment Task   6•3

4. Andréa and Marta are testing three different coolers to see which keeps the coldest temperature. They placed a bag of ice in each cooler, closed the coolers, and then measured the air temperature inside each after 90 minutes. The temperatures are recorded in the table below:

Cooler	A	B	C
Temperature (°C)	−2.91	5.7	−4.3

   Marta wrote the following inequality statement about the temperatures:

   $$-4.3 < -2.91 < 5.7.$$

   Andréa claims that Marta made a mistake in her statement and that the inequality statement should be written as

   $$-2.91 < -4.3 < 5.7.$$

   a. Is either student correct? Explain.

   b. The students want to find a cooler that keeps the temperature inside the cooler more than 3 degrees below the freezing point of water (0°C) after 90 minutes. Indicate which of the tested coolers meets this goal, and explain why.

5. Mary manages a company that has been hired to flatten a plot of land. She took several elevation samples from the land and recorded those elevations below:

Elevation Sample	A	B	C	D	E	F
Elevation (ft. above sea level)	826.5	830.2	832.0	831.1	825.8	827.1

a. The landowner wants the land flat and at the same level as the road that passes in front of it. The road's elevation is 830 feet above sea level. Describe in words how elevation samples B, C, and E compare to the elevation of the road.

b. The table below shows how some other elevation samples compare to the level of the road:

Elevation Sample	G	H	I	J	K	L
Elevation (ft. from the road)	3.1	−0.5	2.2	1.3	−4.5	−0.9

Write the values in the table in order from least to greatest.

c. Indicate which of the values from the table in part (b) is farthest from the elevation of the road. Use absolute value to explain your answer.

Mid-Module Assessment Task 6•3

A Progression Toward Mastery

Assessment Task Item		STEP 1 Missing or incorrect answer and little evidence of reasoning or application of mathematics to solve the problem.	STEP 2 Missing or incorrect answer but evidence of some reasoning or application of mathematics to solve the problem.	STEP 3 A correct answer with some evidence of reasoning or application of mathematics to solve the problem, or an incorrect answer with substantial evidence of solid reasoning or application of mathematics to solve the problem.	STEP 4 A correct answer supported by substantial evidence of solid reasoning or application of mathematics to solve the problem.
1	a 6.NS.C.5 6.NS.C.6a	Student is able to determine that the given water level is below normal (or low water) but does not indicate a clear understanding of zero or the numbers above and below zero on the number line. OR Student is unable to determine that the water level is below normal.	Student correctly states that 0 represents the normal water level but does not clearly describe the meanings of numbers above and below zero on the number line.	Student correctly states that 0 represents the normal water level and that either the numbers above zero represent above normal water levels or the numbers below zero represent below normal water levels but does not clearly describe both.	Student correctly states that 0 represents normal water level, numbers above zero (or positive numbers) represent above normal water levels, and numbers below zero (or negative numbers) represent below normal water levels.
	b 6.NS.C.5	Student response is missing or incomplete. For instance, student makes a general statement that the river's water level is below normal but does not refer to magnitude, direction, or a location on the number line.	Student response is incomplete but shows some evidence of understanding such as stating that the water level is at −2 or −1.9 or that the water level is 2 feet (or 1.9 feet) but without details such as units of measurement or direction.	Student correctly interprets the picture to indicate that the current water level is below normal but states it is *exactly* 2 feet below normal water level rather than *nearly* 2 feet below normal water level.	Student response is correct and complete. Student states that the picture indicates that the river's current water level is *about* 2 feet below normal water level.

Module 3: Rational Numbers

131

	c 6.NS.C.5 6.NS.C.6a	Student answer is incomplete or missing. Explanation shows little or no evidence of how to find opposites on a number line.	Student completes the first step stating that the opposite of −2 is 2 (or the opposite of −1.9 is 1.9, etc.) but with no further details or correct statements.	Student correctly states that the opposite of −2 is 2 (or the opposite of −1.9 is 1.9, etc.) and that in the opposite situation the river's water level would be higher than normal but does not clearly describe its location on the number line/gauge. OR Student correctly identifies and clearly describes the location of the opposite number on the number line/gauge but does not address what this level would mean in the context of the situation.	Student correctly addresses all parts of the question. Student correctly states that the opposite of −2 is 2 (or the opposite of −1.9 is 1.9, etc.) and explains where the positive number is located specifying the number of units above 0 or on the opposite side of zero from the negative value. Student also states that the positive number would mean the river's water level is that many feet higher than the normal level.
	d 6.NS.C.5	Student explanation is missing. OR The written explanation demonstrates little or no correct mathematical interpretation of the situation, such as claiming that tomorrow's water level would be below the current level shown.	Student correctly states that the water level would be higher than the level shown but does not provide a specific reading (level) and does not provide adequate reasoning to support the claim.	Student correctly states that the water would rise to a specific level higher than the level shown and identifies the new level but fails to provide a clear explanation to support the claim. OR Student correctly states that the water level would be higher than the level shown and provides adequate reasoning to support the claim but does not provide a specific reading (level).	Student response is complete and correct. Student states that the water level would rise to a specific level higher than the level shown, identifies a specific new level, and provides a clear explanation to support the claim.
2	a 6.NS.C.5 6.NS.C.6a	Student answer is incorrect or missing. Student neither arrives at both rational number representations nor provides a correct explanation of the difference in the two transactions, although one transaction may have been represented with a correct rational number.	Student correctly explains the difference in the two transactions but makes an error in representing one or both transactions as rational numbers. OR Student uses the correct two rational numbers to represent the transactions but does not explain the difference in the transactions.	Student correctly represents the check as −8.98 and the deposit as 8.98. Student explains that the check decreases the account balance or that the deposit increases the account balance but not both.	Student response is complete and correct. The check is represented as −8.98, and the deposit is represented as 8.98. Student provides a clear and accurate explanation of the difference in the transactions.

Module 3: Rational Numbers

A STORY OF RATIOS — Mid-Module Assessment Task — 6•3

	b 6.NS.C.6a 6.NS.C.6c	Student work is missing or incomplete. Student correctly locates and/or labels one point on the number line but shows no other accurate work.	Student shows intent to graph −8.98 and 8.98 correctly on the number line but does not accurately locate both points and/or has an error in the scale. Student does not state that the numbers are opposites.	Student correctly locates and labels −8.98 and 8.98 on the number line but does not state that the numbers are opposites. OR Student correctly locates −8.98 and 8.98 on the number line and states that the numbers are opposites but does not label the points.	Student response is correct and complete. Student correctly locates and labels −8.98 and 8.98 on the number line and states that the numbers are opposites.
	c 6.NS.C.7c	Student explanation shows little or no evidence of understanding the concept of absolute value.	Student response indicates some evidence of correct reasoning such as stating that the transactions change the account balance by the same amount of money, but the explanation is incomplete and does not include a direct reference to absolute value.	Student states that a debit of $8.98 and credit of $8.98 are similar because both have the same absolute value, but the written response is not complete and does not demonstrate evidence of solid reasoning.	Student response is correct and complete. Student explains that a debit of $8.98 and a credit of $8.98 are similar because the two transactions, which are represented by −8.98 and 8.98, have the same absolute value, which is 8.98; so they change the account balance by the same amount of money but in opposite directions.
3	**a** 6.NS.C.5 6.NS.C.6c	Student accurately locates and labels two of the five points, at most, on the number line using rational numbers.	Student accurately locates and labels three of the five points on the number line using rational numbers.	Student accurately locates and labels four of the five points on the number line using rational numbers.	Student accurately locates and labels all five points on the number line using rational numbers.
	b 6.NS.C.6c 6.NS.C.7c	Student response is incomplete and incorrect, such as stating that a competitor other than Florence is closest to the checkpoint without explaining why.	Student states that Florence is closest to the checkpoint without justification. OR Student states another competitor's name and attempts to justify the answer, but the explanation is incomplete.	Student states that Florence is closest to the checkpoint, but the justification contains an error. OR Student states another competitor's name and justifies the answer based on the response to part (a).	Student correctly states that Florence is closest to the checkpoint and provides clear and accurate justification for the claim.

Module 3: Rational Numbers

A STORY OF RATIOS

Mid-Module Assessment Task 6•3

	c 6.NS.C.5 6.NS.C.6c 6.NS.C.7c	Student response is incomplete and incorrect, such as stating that two competitors *other than* Rebecca and Lita are the same distance from the checkpoint, and no further explanation is provided.	Student is able to determine that Rebecca and Lita are the same distance away from the checkpoint, but the explanation does not address whether or not the competitors are in the same location.	Student states that Rebecca and Lita are the same distance from the checkpoint but on opposite sides; however, the explanation does not specifically answer whether or not the competitors are in the same location.	Student correctly indicates and explains that Rebecca and Lita are both 0.5 mile from the checkpoint but that they are positioned on opposite sides of the checkpoint, and so they are not in the same location.
	d 6.NS.C.7b 6.NS.C.7c	Student explanation shows little or no evidence of understanding. For instance, student incorrectly states Nancy is closer to finishing the race with no explanation why.	Student incorrectly determines Nancy is closer to finishing the race but uses a valid argument based on earlier work. OR Student correctly states that Florence is closer to finishing the race but with no further explanation.	Student correctly states Florence is closer to finishing the race, but the justification for the claim contains an error in reasoning or a misrepresentation.	Student correctly determines and states that Florence is closer to finishing the race and justifies the claim using valid and detailed reasoning.
4	a 6.NS.C.7a 6.NS.C.7b	Student explanation shows little or no evidence of understanding. For instance, student states that neither Marta nor Andréa is correct.	Student states that Marta is correct but does not support the claim. OR Student states that Andréa is correct, but the explanation includes an error in reasoning.	Student correctly states that Marta is correct, but the explanation contains reasoning that is not clear and complete.	Student response is correct and complete. Student states that Marta is correct, justifying the claim by accurately describing the order of the rational numbers on the number line.
	b 6.NS.C.7b 6.NS.C.7d	Student incorrectly states cooler A or B and C met the goal and provides no justification or provides an explanation that contains multiple errors in reasoning.	Student states that cooler C met the goal but provides no justification for the claim. OR Student determines that coolers A and C met the goal and includes a complete explanation but erroneously identifies −2.91 degrees as being more than 3 degrees below zero.	Student correctly states that cooler C met the goal and justifies the claim, but the explanation contains a slight error. For instance, student describes the numbers to the left of −3 on the number line as being more than −3 rather than less than −3.	Student correctly states that cooler C met the goal and justifies the claim by describing that "more than 3 degrees below zero" indicates the numbers must be to the left of −3 (below −3) on the number line and that −4.3 is the only piece of data that meets that criteria.

Module 3:   Rational Numbers

5	a 6.NS.C5 6.NS.C.7b	Student comparison of the elevation samples to the level of the road is incorrect. The written work shows little or no understanding of ordering rational numbers.	Student comparison of the elevation samples to the level of the road is partially correct. Student correctly compares only one or two of the samples (B, C, or E) to the elevation of the road.	Student states that sample C is higher than the elevation of the road and sample E is lower than the elevation of the road and that sample B is about level with the road but *does not distinguish* whether sample B's elevation level is higher or lower than 830 feet.	Student accurately describes each sample's relative position compared to the elevation of the road, stating that samples B and C are higher than the elevation of the road, and sample E is lower than the elevation of the road.
	b 6.NS.C.7b	Student response shows little or no evidence of understanding. Student may place the negative values left of the positive values but makes several errors in order.	Student response shows some evidence of understanding. Student correctly orders four of the six values from least to greatest.	Student correctly orders all six values from least to greatest but copies one of the values incorrectly.	Student correctly orders and writes all six values from least to greatest (i.e., $-4.5 < -0.9 < -0.5 < 1.3 < 2.2 < 3.1$).
	c 6.NS.C.7c	Student indicates sample K but does not provide any further detail. OR Student states a different sample such as G and justifies the choice using clear reasoning but does not address absolute value.	Student indicates sample K and provides a valid explanation but does not address absolute value in the explanation. OR Student incorrectly states sample G and justifies the choice by addressing the order of the positive numbers.	Student correctly states sample K and justifies the statement using absolute value correctly in the explanation, but the explanation is not complete. OR Student incorrectly states sample G and justifies the choice using absolute value in a correct manner but without considering the absolute value of -4.5 for sample K.	Student correctly states sample K is the farthest from the elevation of the road and justifies the statement by comparing the absolute values of the samples from the table in part (b) using the order of rational numbers to reach the answer.

Module 3: Rational Numbers

Mid-Module Assessment Task 6•3

Name _____ Date _____

1. The picture below is a flood gauge that is used to measure how far (in feet) a river's water level is above or below its normal level.

    a. Explain what the number 0 on the gauge represents, and explain what the numbers above and below 0 represent.

    The number 0 represents the normal average water level in the river. The numbers below 0 indicate low water level and the numbers above 0 indicate high water level.

    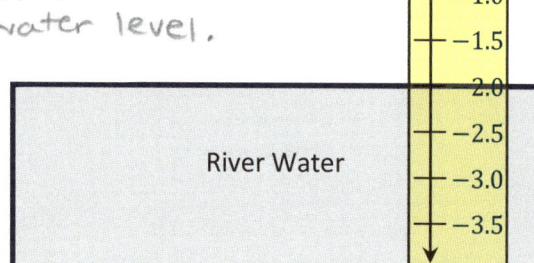

    b. Describe what the picture indicates about the river's current water level.

    The river's water level is about 2 feet below normal.

    c. What number represents the opposite of the water level shown in the picture, and where is it located on the gauge? What would it mean if the river water was at that level?

    The water level is currently at approximately -2.0 feet. The opposite of -2 is 2. 2 is on the opposite side of 0, or above zero. If the river was at 2, the water level would be higher than normal.

    d. If heavy rain is in the forecast for the area for the next 24 hours, what reading might you expect to see on this gauge tomorrow? Explain your reasoning.

    I would expect to see the water level closer to 0 or even higher. Heavy rain should cause the amount of water in the river to increase, so its level would move up the number line.

2. Isaac made a mistake in his checkbook. He wrote a check for $8.98 to rent a video game but mistakenly recorded it in his checkbook as an $8.98 deposit.

    a. Represent each transaction with a rational number, and explain the difference between the transactions.

    A check will decrease his account balance so it can be represented by −8.98.
    A deposit will increase his account balance so it can be represented by 8.98.

    b. On the number line below, locate and label the points that represent the rational numbers listed in part (a). Describe the relationship between these two numbers. Zero on the number line represents Isaac's balance before the mistake was made.

    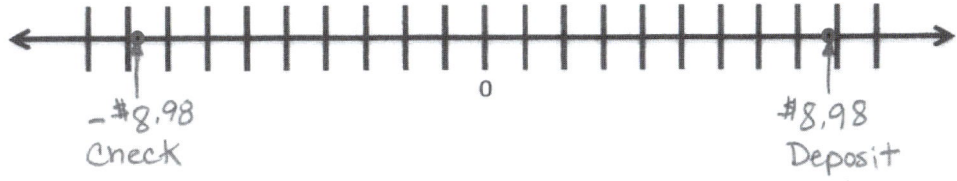

    −$8.98 Check $8.98 Deposit

    The numbers that represent the two transactions are opposites.

    c. Use absolute value to explain how a debit of $8.98 and a credit of $8.98 are similar.

    The check and deposit have the same absolute value (8.98) so they will change his account balance by the same amount of money, but they change the balance in opposite directions.

3. A local park's programs committee is raising money by holding mountain bike races on a course through the park. During each race, a computer tracks the competitors' locations on the course using GPS tracking. The table shows how far each competitor is from a checkpoint.

Number	Competitor Name	Distance to Checkpoint
223	Florence	0.1 miles before
231	Mary	$\frac{2}{5}$ miles past
240	Rebecca	0.5 miles before
249	Lita	$\frac{1}{2}$ miles past
255	Nancy	$\frac{2}{10}$ miles before

a. The checkpoint is represented by 0 on the number line. Locate and label points on the number line for the positions of each listed participant. Label the points using rational numbers.

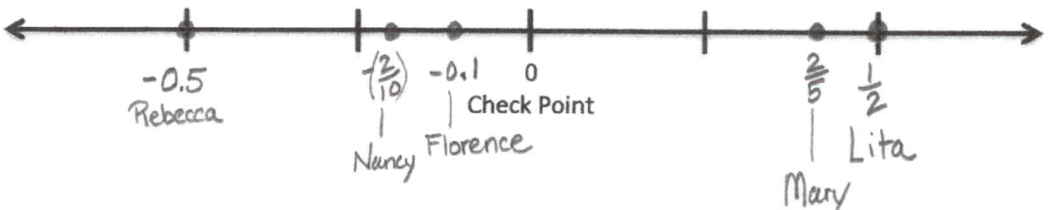

b. Which of the competitors is closest to the checkpoint? Explain.

Florence is closest to the checkpoint because her distance to the checkpoint is 0.1 miles which is less than any of the other girls' distances.

c. Two competitors are the same distance from the checkpoint. Are they in the same location? Explain.

Rebecca and Lita are both 0.5 miles from the checkpoint; they are just on opposite sides of the checkpoint.

d. Who is closer to finishing the race, Nancy or Florence? Support your answer.

Florence is closer to finishing the race because the number representing her position (-0.1) is to the right of (-$\frac{2}{10}$) on the number line which is Nancy's position.

4. Andréa and Marta are testing three different coolers to see which keeps the coldest temperature. They placed a bag of ice in each cooler, closed the coolers, and then measured the air temperature inside each after 90 minutes. The temperatures are recorded in the table below:

Cooler	A	B	C
Temperature (°C)	−2.91	5.7	−4.3

Marta wrote the following inequality statement about the temperatures:

$$-4.3 < -2.91 < 5.7.$$

Andréa claims that Marta made a mistake in her statement and that the inequality statement should be written as

$$-2.91 < -4.3 < 5.7.$$

a. Is either student correct? Explain.

Marta is correct because the order of the numbers in her inequality is the same as the order of the numbers on the number line moving from left to right (or from down to up.).

b. The students want to find a cooler that keeps the temperature inside the cooler more than 3 degrees below the freezing point of water (0°C) after 90 minutes. Indicate which of the tested coolers meets this goal, and explain why.

More than 3 degrees below 0°C means less than −3°C. The only cooler to keep the temperature less than −3°C is cooler C. Cooler C held a temperature of −4.3°C which is to the left of −3°C on the number line.

5. Mary manages a company that has been hired to flatten a plot of land. She took several elevation samples from the land and recorded those elevations below:

Elevation Sample	A	B	C	D	E	F
Elevation (ft. above sea level)	826.5	830.2	832.0	831.1	825.8	827.1

a. The landowner wants the land flat and at the same level as the road that passes in front of it. The road's elevation is 830 feet above sea level. Describe in words how elevation samples B, C, and E compare to the elevation of the road.

Samples B and C are higher than 830 feet and so higher than the road. Sample E is lower than 830 feet and so lower than the road.

b. The table below shows how some other elevation samples compare to the level of the road:

Elevation Sample	G	H	I	J	K	L
Elevation (ft. from the road)	3.1	−0.5	2.2	1.3	−4.5	−0.9

Write the values in the table in order from least to greatest.

−4.5 < −0.9 < −0.5 < 1.3 < 2.2 < 3.1

c. Indicate which of the values from the table in part (b) is farthest from the elevation of the road. Use absolute value to explain your answer.

−4.5 (Sample K) is furthest from the elevation of the road because its absolute value (4.5) is greater than the absolute values of the other samples in the table.

A STORY OF RATIOS

6 GRADE

Mathematics Curriculum

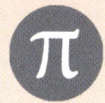

GRADE 6 • MODULE 3

Topic C

Rational Numbers and the Coordinate Plane

6.NS.C.6b, 6.NS.C.6c, 6.NS.C.8

Focus Standards:	6.NS.C.6b 6.NS.C.6c	Understand a rational number as a point on the number line. Extend number line diagrams and coordinate axes familiar from previous grades to represent points on the line and in the plane with negative number coordinates. b. Understand signs of numbers in ordered pairs as indicating locations in quadrants of the coordinate plane; recognize that when two ordered pairs differ only by signs, the locations of the points are related by reflections across one or both axes. c. Find and position integers and other rational numbers on a horizontal or vertical number line diagram; find and position pairs of integers and other rational numbers on a coordinate plane.
	6.NS.C.8	Solve real-world and mathematical problems by graphing points in all four quadrants of the coordinate plane. Include use of coordinates and absolute value to find distances between points with the same first coordinate or the same second coordinate.
Instructional Days:	6	
Lesson 14:	Ordered Pairs (P)[1]	
Lesson 15:	Locating Ordered Pairs on the Coordinate Plane (P)	
Lesson 16:	Symmetry in the Coordinate Plane (P)	
Lesson 17:	Drawing the Coordinate Plane and Points on the Plane (P)	
Lesson 18:	Distance on the Coordinate Plane (P)	
Lesson 19:	Problem Solving and the Coordinate Plane (E)	

In Topic C, students transition from the number line model to represent points in the coordinate plane (**6.NS.C.6c**). Their conceptual understanding of symmetry from Grade 4 and their experience with the first quadrant of the coordinate plane in Grade 5 (**4.G.A.3**, **5.G.A.1**, **5.G.A.2**) serve as a significant foundation as they extend the plane to all four quadrants. In Lesson 14, students use ordered pairs of rational numbers to

---

[1]Lesson Structure Key: **P**-Problem Set Lesson, **M**-Modeling Cycle Lesson, **E**-Exploration Lesson, **S**-Socratic Lesson

Topic C: Rational Numbers and the Coordinate Plane

141

This work is derived from Eureka Math ™ and licensed by Great Minds. ©2015 Great Minds. eureka-math.org

name points on a grid, and given a point's location, they identify the first number in the ordered pair as the first coordinate and the second number as the second coordinate. In Lessons 15–17, students construct the plane; identify the axes, quadrants, and origin; and graph points in the plane, using an appropriate scale on the axes. Students recognize the relationship that exists between points whose coordinates differ only by signs (as reflections across one or both axes) and locate such points using the symmetry in the plane (**6.NS.C.6b**). For instance, they recognize that the points $(3, 4)$ and $(3, -4)$ are both equal distance from the x-axis on the same vertical line, and so the points are reflections in the x-axis. In Lessons 18 and 19, students graph points in the coordinate plane and use absolute value to find the lengths of vertical and horizontal segments to solve real-world problems (**6.NS.C.8**).

Lesson 14: Ordered Pairs

Student Outcomes

- Students use ordered pairs to name points in a grid and to locate points on a map.
- Students identify the first number in an ordered pair as the *first coordinate* and the second number as the *second coordinate*.

Lesson Notes

Students understand the use of ordered pairs of numbers as describing the locations of points on a plane in various situations. They recognize the significance of the order of numbers in ordered pairs by looking at the different interpretations.

Before starting the lesson, the teacher must make copies of the scenarios in Example 2 and cut them apart in order to provide each group with exactly one scenario.

Classwork

Opening Exercise (5 minutes)

Before students arrive, arrange their desks into straight rows. Assign a number (1, 2, 3, ...) to each row and also to the seats in each row starting at the front with seat 1. As students enter the room, give them a sticky note containing a pair of numbers corresponding with the seating locations in the room. Instruct students to find the seat described by their sticky note, apply the sticky note to the desk, and be seated.

Most students are confused as only those with matching numbers are able to find their seats. Monitor conversations taking place between students as they agree upon a convention (e.g., that the first number represents the row, and the second number represents the seat).

- How did you find your seat in the classroom? (Answers will vary.)
- Did the order of the numbers matter? Why or why not?
  - *The order mattered since there are two different seats that involve the numbers 2 and 3. For instance, row 2, seat 3, and row 3, seat 2.*

Example 1 (5 minutes): The *Order* in Ordered Pairs

Instruct students to rotate their desks 90 degrees in one direction. This changes the orientation of the rows so that students can better see the meanings of each of the coordinates. Students understand that the coordinates of their location from the Opening Exercise (in most cases) are different in Example 1. For example, the student sitting in row 1, seat 3, for the Opening Exercise, is now sitting in row 3, seat 1.

A STORY OF RATIOS　　　　　　　　　　　　　　　　　　　　　　　　　　　　Lesson 14　6•3

> **Example 1: The *Order* in Ordered Pairs**
>
> The first number of an ordered pair is called the ___first coordinate___.
>
> The second number of an ordered pair is called the ___second coordinate___.

Define the first and second coordinates in this example as (row number, seat number). Ask all students in the classroom to stand. Call out an appropriate ordered pair, and ask for the student in that location to raise his hand, briefly explain why the ordered pair of numbers describes that student's position in the room, and then be seated. Now have that student call out a different ordered pair that corresponds with the location of another student. Continue this process until all students have participated.

Example 2 (10 minutes): Using Ordered Pairs to Name Locations

Divide students into small groups, and provide each group with one of the ordered pair scenarios below. Students read their scenarios and describe how the ordered pair is being used, indicating what defines the first coordinate and what defines the second coordinate. Allow groups 5 minutes to read and discuss the scenario and prepare a response to report out to the class.

Scaffolding:
Provide extra practice in naming locations using ordered pairs by playing a game on the coordinate plane where students try to guess the locations of their opponents' points.

> **Example 2: Using Ordered Pairs to Name Locations**
>
> Describe how the ordered pair is being used in your scenario. Indicate what defines the first coordinate and what defines the second coordinate in your scenario.

Ordered pairs are like a set of directions; they indicate where to go in one direction and then indicate where to go in the second direction.

- Scenario 1: The seats in a college football stadium are arranged into 210 sections, with 144 seats in each section. Your ticket to the game indicates the location of your seat using the ordered pair of numbers (123, 37). Describe the meaning of each number in the ordered pair and how you would use them to find your seat.

- Scenario 2: Airline pilots use measurements of longitude and latitude to determine their location and to find airports around the world. Longitude is measured as 0–180° east or 0–180° west of a line stretching from the North Pole to the South Pole through Greenwich, England, called the *prime meridian*. Latitude is measured as 0–90° north or 0–90° south of the earth's equator. A pilot has the ordered pair (90° west, 30° north). What does each number in the ordered pair describe? How would the pilot locate the airport on a map? Would there be any confusion if a pilot were given the ordered pair (90°, 30°)? Explain.

- Scenario 3: Each room in a school building is named by an ordered pair of numbers that indicates the number of the floor on which the room lies, followed by the sequential number of the room on the floor from the main staircase. A new student at the school is trying to get to science class in room 4–13. Describe to the student what each number means and how she should use the number to find her classroom. Suppose there are classrooms below the main floor. How might these rooms be described?

Ask student groups to report their answers to the scenarios aloud to the class.

144　　Lesson 14:　Ordered Pairs

A STORY OF RATIOS   Lesson 14   6•3

Exercises 1–2 (12 minutes)

Students use the gridded maps in the student materials to name points that correspond with the given ordered pairs (and vice versa). The first coordinates represent numbers on the line labeled x, and the second coordinates represent numbers on the line labeled y.

Scaffolding:
If students do not understand the negative numbers on the vertical axis, review with students how the floors below ground level might be described in Scenario 3 from Example 2.

Exercises

The first coordinates of the ordered pairs represent the numbers on the line labeled x, and the second coordinates represent the numbers on the line labeled y.

1. Name the letter from the grid below that corresponds with each ordered pair of numbers below.

   a. $(1, 4)$ b. $(0, 5)$
       Point F   **Point A**

   c. $(4, 1)$ d. $(8.5, 8)$
       Point B   **Point L**

   e. $(5, -2)$ f. $(5, 4.2)$
       Point G   **Point H**

   g. $(2, -1)$ h. $(0, 9)$
       Point C   **Point E**

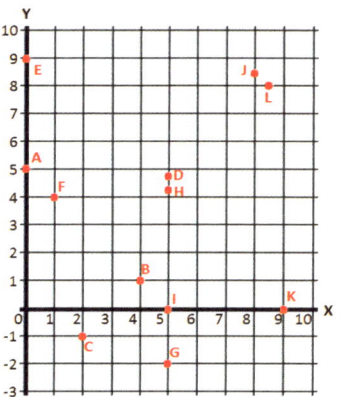

2. List the ordered pair of numbers that corresponds with each letter from the grid below.

   a. Point M b. Point S
       $(5, 7)$ $(-2, 3)$

   c. Point N d. Point T
       $(6, 0)$ $(-3, 2)$

   e. Point P f. Point U
       $(0, 6)$ $(7, 5)$

   g. Point Q h. Point V
       $(2, 3)$ $(-1, 6)$

   i. Point R
       $(0, 3)$

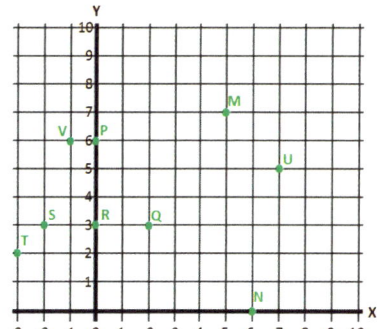

Have students provide the correct answers to the exercises.

Lesson 14:   Ordered Pairs   145

Closing (5 minutes)

- Why does order matter when using ordered pairs of numbers?
  - *The order is important because it provides one specific location in the coordinate plane.*
- Alayna says the order in which the values are given in an ordered pair does not always matter. Give an example of when the order does matter and an example of when the order does not matter.
  - *The order does not matter if the first and second coordinates are the same number. For example, $(3, 3)$ is the same location in the coordinate plane no matter which point is used as the first coordinate. However, order does matter when the two coordinates are not the same. For example, $(1, 3)$ has a different location in the coordinate plane than $(3, 1)$.*
- Explain how to locate points when pairs of integers are used.
  - *The first coordinate describes the location of the point using the horizontal direction. Positive integers indicate moving to the right from zero, and we would move left for negative numbers. The second coordinate describes the location of the point using the vertical direction. Positive numbers indicate moving up from zero, and we would move down from zero for negative integers.*

Lesson Summary

- The order of numbers in an ordered pair is important because the ordered pair should describe <u>one</u> location in the coordinate plane.
- The first number (called the *first coordinate*) describes a location using the horizontal direction.
- The second number (called the *second coordinate*) describes a location using the vertical direction.

Exit Ticket (8 minutes)

A STORY OF RATIOS   Lesson 14 6•3

Name _____   Date _____

Lesson 14: Ordered Pairs

Exit Ticket

1. On the map below, the fire department and the hospital have one matching coordinate. Determine the proper order of the ordered pairs in the map, and write the correct ordered pairs for the locations of the fire department and hospital. Indicate which of their coordinates are the same.

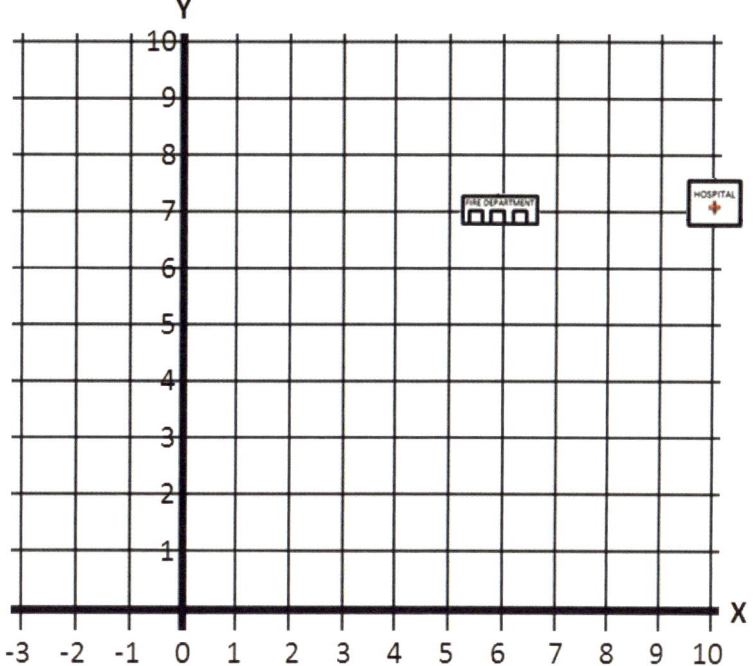

2. On the map above, locate and label the locations of each description below:

   a. The local bank has the same first coordinate as the fire department, but its second coordinate is half of the fire department's second coordinate. What ordered pair describes the location of the bank? Locate and label the bank on the map using point B.

   b. The Village Police Department has the same second coordinate as the bank, but its first coordinate is -2. What ordered pair describes the location of the Village Police Department? Locate and label the Village Police Department on the map using point P.

Lesson 14:   Ordered Pairs   147

A STORY OF RATIOS — Lesson 14 6•3

Exit Ticket Sample Solutions

1. On the map below, the fire department and the hospital have one matching coordinate. Determine the proper order of the ordered pairs in the map, and write the correct ordered pairs for the locations of the fire department and hospital. Indicate which of their coordinates are the same.

   The order of the numbers is (x, y); fire department: $(6, 7)$ and hospital: $(10, 7)$; they have the same second coordinate.

   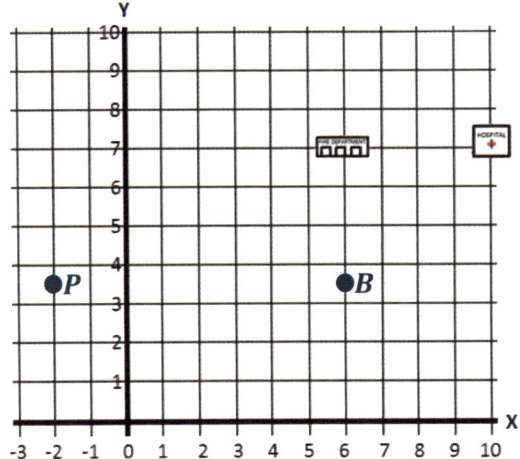

2. On the map above, locate and label the location of each description below:

   a. The local bank has the same first coordinate as the fire department, but its second coordinate is half of the fire department's second coordinate. What ordered pair describes the location of the bank? Locate and label the bank on the map using point B.

      $(6, 3.5)$; see the map image for the correct location of point B.

   b. The Village Police Department has the same second coordinate as the bank, but its first coordinate is -2. What ordered pair describes the location of the Village Police Department? Locate and label the Village Police Department on the map using point P.

      $(-2, 3.5)$; see the map image for the correct location of point P.

Problem Set Sample Solutions

1. Use the set of ordered pairs below to answer each question.

   $$\{(4, 20), (8, 4), (2, 3), (15, 3), (6, 15), (6, 30), (1, 5), (6, 18), (0, 3)\}$$

   a. Write the ordered pair(s) whose first and second coordinate have a greatest common factor of 3.

      $(15, 3)$ and $(6, 15)$

   b. Write the ordered pair(s) whose first coordinate is a factor of its second coordinate.

      $(4, 20), (6, 30), (1, 5),$ and $(6, 18)$

Lesson 14: Ordered Pairs

c. Write the ordered pair(s) whose second coordinate is a prime number.

$(2, 3)$, $(15, 3)$, $(1, 5)$, and $(0, 3)$

2. Write ordered pairs that represent the location of points A, B, C, and D, where the first coordinate represents the horizontal direction, and the second coordinate represents the vertical direction.

$A\ (4, 1);\ B\ (1, -3);\ C\ (6, 0);\ D\ (1, 4)$

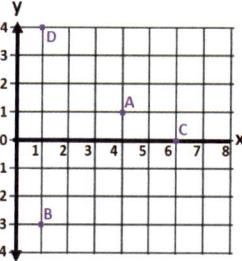

Extension:

3. Write ordered pairs of integers that satisfy the criteria in each part below. Remember that the origin is the point whose coordinates are $(0, 0)$. When possible, give ordered pairs such that (i) both coordinates are positive, (ii) both coordinates are negative, and (iii) the coordinates have opposite signs in either order.

   a. These points' vertical distance from the origin is twice their horizontal distance.

   Answers will vary; examples are $(5, 10)$, $(-2, 4)$, $(-5, -10)$, $(2, -4)$.

   b. These points' horizontal distance from the origin is two units more than the vertical distance.

   Answers will vary; examples are $(3, 1)$, $(-3, 1)$, $(-3, -1)$, $(3, -1)$.

   c. These points' horizontal and vertical distances from the origin are equal, but only one coordinate is positive.

   Answers will vary; examples are $(3, -3)$, $(-8, 8)$.

Lesson 14: Ordered Pairs

Lesson 15: Locating Ordered Pairs on the Coordinate Plane

Student Outcomes

- Students extend their understanding of the coordinate plane to include all four quadrants and recognize that the axes (identified as the x-axis and y-axis) of the coordinate plane divide the plane into four regions called *quadrants* (that are labeled from first to fourth and are denoted by roman numerals).
- Students identify the origin and locate points other than the origin, which lie on an axis.
- Students locate points in the coordinate plane that correspond to given ordered pairs of integers and other rational numbers.

Classwork

Opening Exercise (6 minutes)

Hang posters on the wall, each containing one of the following terms: x-axis, y-axis, x-coordinate, y-coordinate, origin, and coordinate pair. Pair students up, and have them discuss these vocabulary terms and what they remember about the terms from Grade 5. Student pairs then write what they discussed on the posters with the appropriate vocabulary term. Some important aspects for students to remember include the following:

- The x-axis is a horizontal number line; the y-axis is a vertical number line.
- The axes meet forming a 90° angle at the point $(0, 0)$ called the *origin*.

Example 1 (8 minutes): Extending the Axes Beyond Zero

Students recognize that the axes are number lines and, using straightedges, extend the axes on the coordinate plane to include negative numbers revealing the second, third, and fourth quadrants.

- Describe the x-axis. Considering what we have seen in this module, what types of numbers should it include? Why?
  - *The x-axis is a horizontal number line that includes positive and negative numbers. The axis extends in both directions (left and right of zero) because signed numbers represent values or quantities that have opposite directions.*

Have student use their prior knowledge to complete Example 1 individually.

> **Example 1: Extending the Axes Beyond Zero**
>
> The point below represents zero on the number line. Draw a number line to the right starting at zero. Then, follow directions as provided by the teacher.
>
>

- Use a straightedge to extend the x-axis to the left of zero to represent the real number line horizontally, and complete the number line using the same scale as on the right side of zero.

A STORY OF RATIOS   Lesson 15 6•3

- Describe the y-axis. What types of numbers should it include?
  - *The y-axis is a vertical number line that includes numbers on both sides of zero (above and below), and so it includes both positive and negative numbers.*
- Use a straightedge to draw a vertical number line above zero.

Provide students with time to draw.

- Extend the y-axis below zero to represent the real number line vertically, and complete the number line using the same scale as above zero.

Example 2 (4 minutes): Components of the Coordinate Plane

Students examine how to use the axes and the origin of the coordinate plane to determine other locations in the plane.

> **Example 2: Components of the Coordinate Plane**
>
> All points on the coordinate plane are described with reference to the origin. What is the origin, and what are its coordinates?
>
> *The origin is the point where the x- and y-axes intersect. The coordinates of the origin are $(0, 0)$.*

Scaffolding:
- *The term origin means starting point.*
- *A person's country of origin is the country from which he came.*
- *When using a global positioning unit (GPS) when traveling, the origin is where the trip began.*

- The axes of the coordinate plane intersect at their zero coordinates, which is a point called the *origin*. The origin is the reference point from which all points in the coordinate plane are described.

> To describe locations of points in the coordinate plane, we use ___ordered pairs___ of numbers. Order is important, so on the coordinate plane, we use the form (_x y_). The first coordinate represents the point's location from zero on the __x__-axis, and the second coordinate represents the point's location from zero on the __y__-axis.

Exercises 1–3 (8 minutes)

Students locate and label points that lie on the axes of the coordinate plane.

> **Exercises 1–3**
>
> 1. Use the coordinate plane below to answer parts (a)–(c).
>   a. Graph at least five points on the x-axis, and label their coordinates.
>
>   *Points will vary.*
>
>   b. What do the coordinates of your points have in common?
>
>   *Each point has a y-coordinate of 0.*
>
>   c. What must be true about any point that lies on the x-axis? Explain.
>
>   *If a point lies on the x-axis, its y-coordinate must be 0 because the point is located 0 units above or below the x-axis. The x-axis intersects the y-axis at 0.*

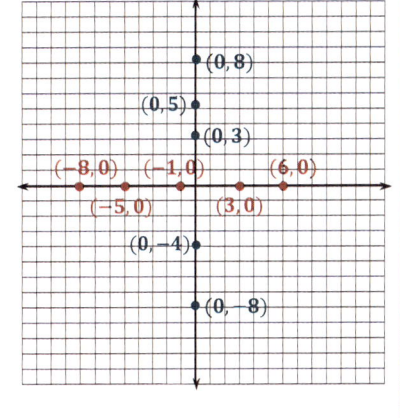

Lesson 15:   Locating Ordered Pairs on the Coordinate Plane   151

This work is derived from Eureka Math ™ and licensed by Great Minds. ©2015 Great Minds. eureka-math.org
G6-M3-TE-B3-1.3.0-07.2015

A STORY OF RATIOS Lesson 15 6•3

MP.7

2. Use the coordinate plane to answer parts (a)–(c).

   a. Graph at least five points on the y-axis, and label their coordinates.

      Points will vary.

   b. What do the coordinates of your points have in common?

      Each point has an x-coordinate of 0.

   c. What must be true about any point that lies on the y-axis? Explain.

      If a point lies on the y-axis, its x-coordinate must be 0 because the point is located 0 units left or right of the y-axis. The y-axis intersects 0 on the x-axis.

3. If the origin is the only point with 0 for both coordinates, what must be true about the origin?

   The origin is the only point that is on both the x-axis and the y-axis.

Example 3 (6 minutes): Quadrants of the Coordinate Plane

Students examine the four regions of the coordinate plane cut by the intersecting axes.

Example 3: Quadrants of the Coordinate Plane

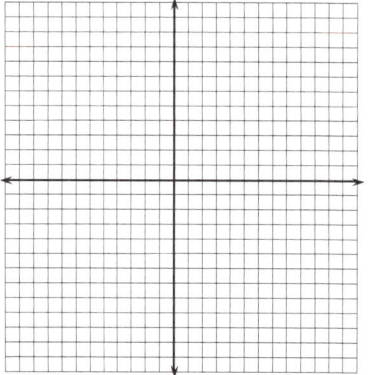

Scaffolding:
- *Remind students that the prefix quad means four, and cite some other examples where the prefix is used.*
- *Some students may not have knowledge of roman numerals. Create a table in which students can compare the standard symbols 1–8 and the roman numerals I–VIII.*

- The x- and y-axes divide the coordinate plane into regions called quadrants. Why are the regions called quadrants?
  - *The axes cut the plane into four regions. The prefix "quad" means four.*
- Which of the four regions did you work with most in Grade 5, and why was it the only region you used?
  - *The region on the top right of the coordinate plane. We only used this region because we had not learned about negative numbers yet.*
- The four quadrants are numbered one through four using roman numerals. The upper-right quadrant is Quadrant I, and the remaining quadrants are numbered moving counterclockwise from Quadrant I: Quadrant II, Quadrant III, and Quadrant IV. What was the first axis that we extended in Example 1, and what did it reveal?
  - *We extended the x-axis to the left beyond zero, and it revealed another region of the coordinate plane.*

152 Lesson 15: Locating Ordered Pairs on the Coordinate Plane

- This top left region is called Quadrant II. Label Quadrant II in your student materials. These regions only make up half of the coordinate plane. Where does the remaining half of the coordinate plane come from? Explain.
  - *We need to extend the y-axis down below zero to show its negative values. This reveals two other regions on the plane, one to the left of the y-axis and one to the right of the y-axis.*
- The quadrants of the coordinate plane are in a counterclockwise direction starting with Quadrant I. Label the remaining quadrants in your student materials.

Exercises 4–6 (5 minutes)

Students locate and label points that lie on the coordinate plane and indicate in which of the four quadrants the points lie.

Exercises 4–6

4. Locate and label each point described by the ordered pairs below. Indicate which of the quadrants the points lie in.

   a. $(7, 2)$

   Quadrant I

   b. $(3, -4)$

   Quadrant IV

   c. $(1, -5)$

   Quadrant IV

   d. $(-3, 8)$

   Quadrant II

   e. $(-2, -1)$

   Quadrant III

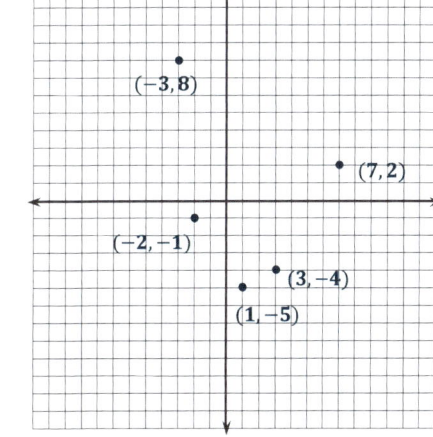

5. Write the coordinates of at least one other point in each of the four quadrants.

   a. Quadrant I

   Answers will vary, but both numbers must be positive.

   b. Quadrant II

   Answers will vary, but the x-coordinate must be negative, and the y-coordinate must be positive.

   c. Quadrant III

   Answers will vary, but both numbers must be negative.

   d. Quadrant IV

   Answers will vary, but the x-coordinate must be positive, and the y-coordinate must be negative.

6. Do you see any similarities in the points within each quadrant? Explain your reasoning.

The ordered pairs describing the points in Quadrant I contain both positive values. The ordered pairs describing the points in Quadrant III contain both negative values. The first coordinates of the ordered pairs describing the points in Quadrant II are negative values, but their second coordinates are positive values. The first coordinates of the ordered pairs describing the points in Quadrant IV are positive values, but their second coordinates are negative values.

Closing (4 minutes)

- If a point lies on an axis, what must be true about its coordinates? Specifically, what is true for a point that lies on the x-axis? The y-axis?
  - *The y-coordinate is always 0 if a point lies on the x-axis. The x-coordinate is always 0 if a point lies on the y-axis.*
- What do you know about the location of a point on the coordinate plane if:
  - Both coordinates are positive?
    - *If both coordinates are positive, the point must be located in Quadrant I.*
  - Only one coordinate is positive?
    - *If only one coordinate is positive, the point is either in Quadrant II or Quadrant IV. If only the first coordinate is positive, then the point is in Quadrant IV. If only the second coordinate is positive, then the point is in Quadrant II.*
  - Both coordinates are negative?
    - *If both coordinates are negative, the point is located in Quadrant III.*
  - One coordinate is zero?
    - *If one coordinate is zero, then the point is located on the x-axis or the y-axis.*
  - Both coordinates are zero?
    - *If both coordinates are zero, then the point represents the origin.*

Lesson Summary

- The x-axis and y-axis of the coordinate plane are number lines that intersect at zero on each number line.
- The axes partition the coordinate plane into four quadrants.
- Points in the coordinate plane lie either on an axis or in one of the four quadrants.

Exit Ticket (4 minutes)

A STORY OF RATIOS   Lesson 15   6•3

Name _____ Date _____

Lesson 15: Locating Ordered Pairs on the Coordinate Plane

Exit Ticket

1. Label the second quadrant on the coordinate plane, and then answer the following questions:

    a. Write the coordinates of one point that lies in the second quadrant of the coordinate plane.

    b. What must be true about the coordinates of any point that lies in the second quadrant?

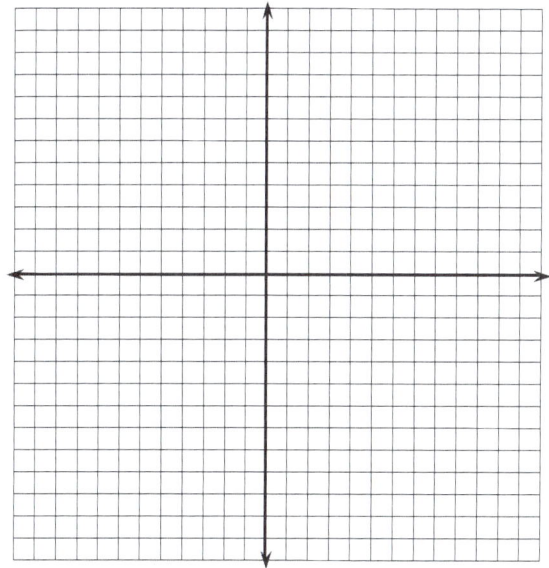

2. Label the third quadrant on the coordinate plane, and then answer the following questions:

    a. Write the coordinates of one point that lies in the third quadrant of the coordinate plane.

    b. What must be true about the coordinates of any point that lies in the third quadrant?

3. An ordered pair has coordinates that have the same sign. In which quadrant(s) could the point lie? Explain.

4. Another ordered pair has coordinates that are opposites. In which quadrant(s) could the point lie? Explain.

A STORY OF RATIOS Lesson 15 6•3

Exit Ticket Sample Solutions

1. Label the second quadrant on the coordinate plane, and then answer the following questions:

   a. Write the coordinates of one point that lies in the second quadrant of the coordinate plane.

   Answers will vary.

   b. What must be true about the coordinates of any point that lies in the second quadrant?

   The x-coordinate must be a negative value, and the y-coordinate must be a positive value.

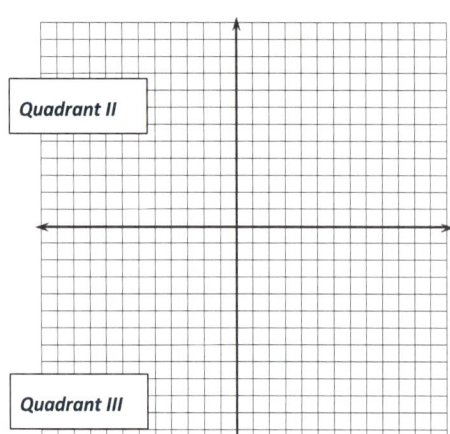

2. Label the third quadrant on the coordinate plane, and then answer the following questions:

   a. Write the coordinates of one point that lies in the third quadrant of the coordinate plane.

   Answers will vary.

   b. What must be true about the coordinates of any point that lies in the third quadrant?

   The x- and y-coordinates of any point in the third quadrant must both be negative values.

3. An ordered pair has coordinates that have the same sign. In which quadrant(s) could the point lie? Explain.

   The point would have to be located either in Quadrant I where both coordinates are positive values or in Quadrant III where both coordinates are negative values.

4. Another ordered pair has coordinates that are opposites. In which quadrant(s) could the point lie? Explain.

   The point would have to be located in either Quadrant II or Quadrant IV because those are the two quadrants where the coordinates have opposite signs. The point could also be located at the origin (0, 0) since zero is its own opposite.

Problem Set Sample Solutions

1. Name the quadrant in which each of the points lies. If the point does not lie in a quadrant, specify which axis the point lies on.

   a. $(-2, 5)$

   Quadrant II

   b. $(8, -4)$

   Quadrant IV

   c. $(-1, -8)$

   Quadrant III

   d. $(9.2, 7)$

   Quadrant I

   e. $(0, -4)$

   None; the point is not in a quadrant because it lies on the y-axis.

156   Lesson 15:   Locating Ordered Pairs on the Coordinate Plane

2. Jackie claims that points with the same x- and y-coordinates must lie in Quadrant I or Quadrant III. Do you agree or disagree? Explain your answer.

   Disagree; most points with the same x- and y-coordinates lie in Quadrant I or Quadrant III, but the origin $(0, 0)$ is on the x- and y-axes, not in any quadrant.

3. Locate and label each set of points on the coordinate plane. Describe similarities of the ordered pairs in each set, and describe the points on the plane.

   a. $\{(-2, 5), (-2, 2), (-2, 7), (-2, -3), (-2, -0.8)\}$

   The ordered pairs all have x-coordinates of -2, and the points lie along a vertical line above and below -2 on the x-axis.

   b. $\{(-9, 9), (-4, 4), (-2, 2), (1, -1), (3, -3), (0, 0)\}$

   The ordered pairs each have opposite values for their x- and y-coordinates. The points in the plane line up diagonally through Quadrant II, the origin, and Quadrant IV.

   c. $\{(-7, -8), (5, -8), (0, -8), (10, -8), (-3, -8)\}$

   The ordered pairs all have y-coordinates of -8, and the points lie along a horizontal line to the left and right of -8 on the y-axis.

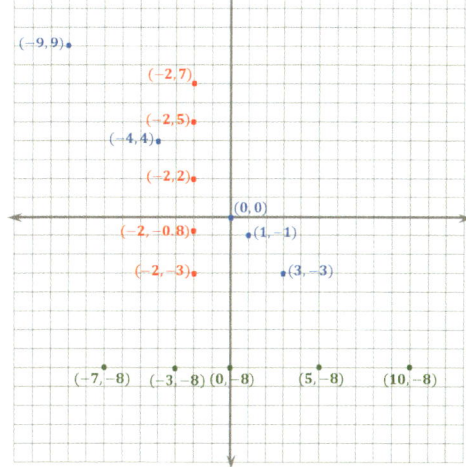

4. Locate and label at least five points on the coordinate plane that have an x-coordinate of 6.

   a. What is true of the y-coordinates below the x-axis?

   The y-coordinates are all negative values.

   b. What is true of the y-coordinates above the x-axis?

   The y-coordinates are all positive values.

   c. What must be true of the y-coordinates on the x-axis?

   The y-coordinates on the x-axis must be 0.

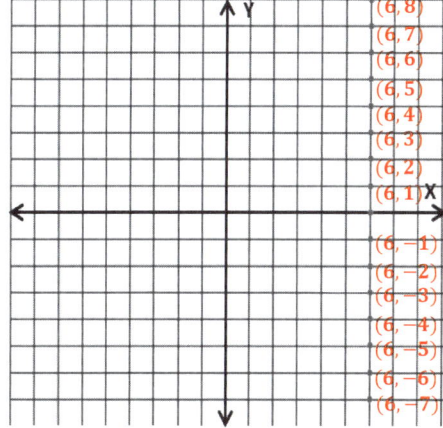

Lesson 15: Locating Ordered Pairs on the Coordinate Plane

Lesson 16: Symmetry in the Coordinate Plane

Student Outcomes

- Students understand that two numbers are said to differ only by signs if they are opposites of each other.
- Students recognize that when two ordered pairs differ only by the sign of one or both of the coordinates, then the locations of the points are related by reflections across one or both axes.

Classwork

Opening Exercise (3 minutes)

> **Opening Exercise**
>
> Give an example of two opposite numbers, and describe where the numbers lie on the number line. How are opposite numbers similar, and how are they different?
>
> *Answers may vary. 2 and −2 are opposites because they are both 2 units from zero on a number line but in opposite directions. Opposites are similar because they have the same absolute value, but they are different because opposites are on opposite sides of zero.*

Example 1 (14 minutes): Extending Opposite Numbers to the Coordinate Plane

Students locate and label points whose ordered pairs differ only by the sign of one or both coordinates. Together, students and the teacher examine the relationships of the points on the coordinate plane and express these relationships in a graphic organizer.

- Locate and label the points $(3, 4)$ and $(-3, 4)$.
- Record observations in the first column of the graphic organizer.

MP.8 The first column of the graphic organizer is teacher led so that students can pay particular attention to the absolute values of coordinates and the general locations of the corresponding points with regard to each axis. Columns 2 and 3 are primarily student led.

- Locate and label the point $(3, -4)$.
- Record observations in the second column of the graphic organizer.
- Locate and label the point $(-3, -4)$.
- Record observations in the third column of the graphic organizer.

A STORY OF RATIOS　　Lesson 16　6•3

Example 1: Extending Opposite Numbers to the Coordinate Plane

Extending Opposite Numbers to the Coordinates of
Points on the Coordinate Plane

Locate and label your points on the coordinate plane to the right. For each given pair of points in the table below, record your observations and conjectures in the appropriate cell. Pay attention to the absolute values of the coordinates and where the points lie in reference to each axis.

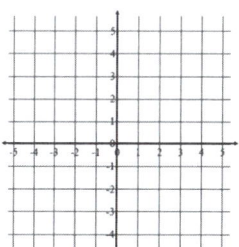

	$(3, 4)$ and $(-3, 4)$	$(3, 4)$ and $(3, -4)$	$(3, 4)$ and $(-3, -4)$
Similarities of Coordinates	*Same y-coordinates* *The x-coordinates have the same absolute value.*	*Same x-coordinates* *The y-coordinates have the same absolute value.*	*The x-coordinates have the same absolute value.* *The y-coordinates have the same absolute value.*
Differences of Coordinates	*The x-coordinates are opposite numbers.*	*The y-coordinates are opposite numbers.*	*Both the x- and y-coordinates are opposite numbers.*
Similarities in Location	*Both points are 4 units above the x-axis and 3 units away from the y-axis.*	*Both points are 3 units to the right of the y-axis and 4 units away from the x-axis.*	*Both points are 3 units from the y-axis and 4 units from the x-axis.*
Differences in Location	*One point is 3 units to the right of the y-axis; the other is 3 units to the left of the y-axis.*	*One point is 4 units above the x-axis; the other is 4 units below.*	*One point is 3 units right of the y-axis; the other is 3 units left. One point is 4 units above the x-axis; the other is 4 units below.*
Relationship Between Coordinates and Location on the Plane	*The x-coordinates are opposite numbers, so the points lie on opposite sides of the y-axis. Because opposites have the same absolute value, both points lie the same distance from the y-axis. The points lie the same distance above the x-axis, so the points are symmetric about the y-axis. A reflection across the y-axis takes one point to the other.*	*The y-coordinates are opposite numbers, so the points lie on opposite sides of the x-axis. Because opposites have the same absolute value, both points lie the same distance from the x-axis. The points lie the same distance right of the y-axis, so the points are symmetric about the x-axis. A reflection across the x-axis takes one point to the other.*	*The points have opposite numbers for x- and y-coordinates, so the points must lie on opposite sides of each axis. Because the numbers are opposites and opposites have the same absolute values, each point must be the same distance from each axis. A reflection across one axis followed by a reflection across the other axis takes one point to the other.*

Lesson 16: Symmetry in the Coordinate Plane

A STORY OF RATIOS Lesson 16 6•3

Exercises 1–2 (5 minutes)

Exercises

In each column, write the coordinates of the points that are related to the given point by the criteria listed in the first column of the table. Point $S(5,3)$ has been reflected over the x- and y-axes for you as a guide, and its images are shown on the coordinate plane. Use the coordinate grid to help you locate each point and its corresponding coordinates.

Given Point:	$S(5,3)$	$(-2,4)$	$(3,-2)$	$(-1,-5)$
The given point is reflected across the x-axis.	$M(5,-3)$	$(-2,-4)$	$(3,2)$	$(-1,5)$
The given point is reflected across the y-axis.	$L(-5,3)$	$(2,4)$	$(-3,-2)$	$(1,-5)$
The given point is reflected first across the x-axis and then across the y-axis.	$A(-5,-3)$	$(2,-4)$	$(-3,2)$	$(1,5)$
The given point is reflected first across the y-axis and then across the x-axis.	$A(-5,-3)$	$(2,-4)$	$(-3,2)$	$(1,5)$

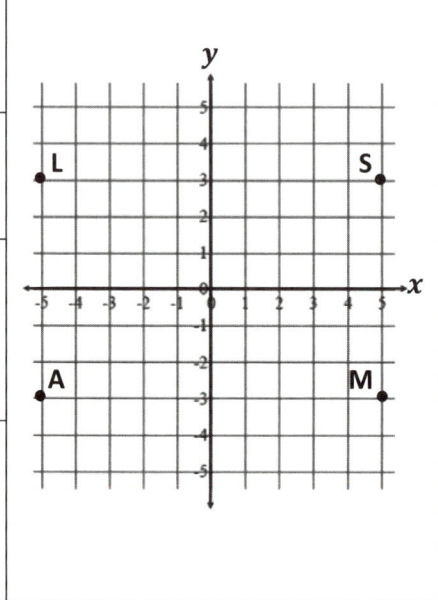

1. When the coordinates of two points are (x,y) and $(-x,y)$, what line of symmetry do the points share? Explain.

   They share the y-axis because the y-coordinates are the same and the x-coordinates are opposites, which means the points will be the same distance from the y-axis but on opposite sides.

2. When the coordinates of two points are (x,y) and $(x,-y)$, what line of symmetry do the points share? Explain.

   They share the x-axis because the x-coordinates are the same and the y-coordinates are opposites, which means the points will be the same distance from the x-axis but on opposite sides.

Example 2 (8 minutes): Navigating the Coordinate Plane Using Reflections

Have students use a pencil eraser or finger to navigate the coordinate plane given verbal prompts. Then, circulate the room during the example to assess students' understanding and provide assistance as needed.

Scaffolding:
Project each prompt so that visual learners can follow along with the steps.

- Begin at $(7,2)$. Move 3 units down, and then reflect over the y-axis. Where are you?
  - $(-7,-1)$

Lesson 16: Symmetry in the Coordinate Plane

A STORY OF RATIOS   **Lesson 16**   **6•3**

- Begin at $(4, -5)$. Reflect over the x-axis, and then move 7 units down and then to the right 2 units. Where are you?
    - $(6, -2)$
- Begin at $(-3, 0)$. Reflect over the x-axis, and then move 6 units to the right. Move up two units, and then reflect over the x-axis again. Where are you?
    - $(3, -2)$
- Begin at $(-2, 8)$. Decrease the y-coordinate by 6 units. Reflect over the y-axis, and then move down 3 units. Where are you?
    - $(2, -1)$
- Begin at $(5, -1)$. Reflect over the x-axis, and then reflect over the y-axis. Where are you?
    - $(-5, 1)$

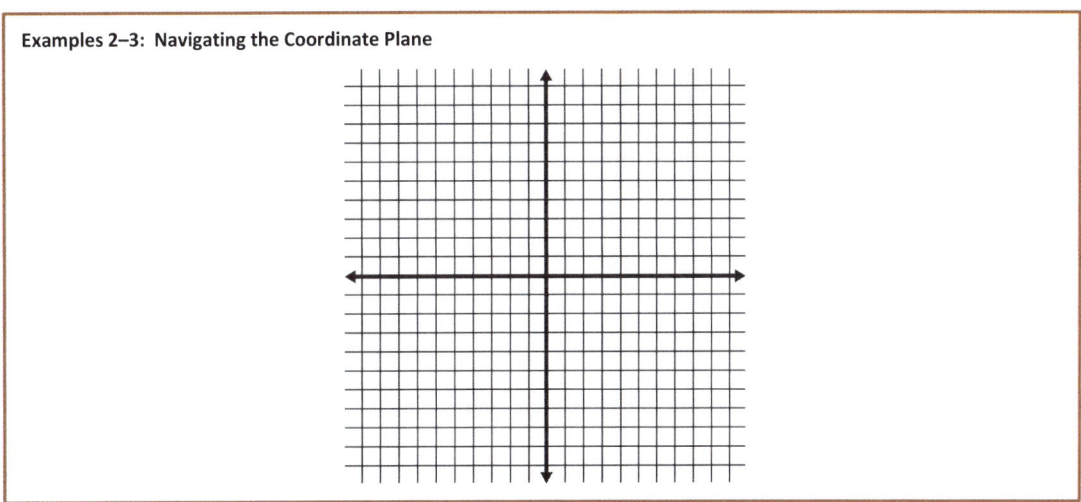

Examples 2–3: Navigating the Coordinate Plane

Example 3 (7 minutes): Describing How to Navigate the Coordinate Plane

Given a starting point and an ending point, students describe a sequence of directions using at least one reflection about an axis to navigate from the starting point to the ending point. Once students have found a sequence, have them find another sequence while classmates finish the task.

- Begin at $(9, -3)$, and end at $(-4, -3)$. Use exactly one reflection.
    - *Possible answer: Reflect over the y-axis, and then move 5 units to the right.*
- Begin at $(0, 0)$, and end at $(5, -1)$. Use exactly one reflection.
    - *Possible answer: Move 5 units right, 1 unit up, and then reflect over the x-axis.*
- Begin at $(0, 0)$, and end at $(-1, -6)$. Use exactly two reflections.
    - *Possible answer: Move right 1 unit, reflect over the y-axis, up 6 units, and then reflect over the x-axis.*

Lesson 16:   Symmetry in the Coordinate Plane   161

Closing (4 minutes)

- When the coordinates of two points differ only by one sign, such as $(-8, 2)$ and $(8, 2)$, what do the similarities and differences in the coordinates tell us about their relative locations on the plane?
  - *The y-coordinates are the same for both points, which means the points are on the same horizontal line. The x-coordinates differ because they are opposites, which means the points are symmetric across the y-axis.*
- What is the relationship between $(5, 1)$ and $(5, -1)$? Given one point, how can you locate the other?
  - *If you start at either point and reflect over the x-axis, you will end at the other point.*

Exit Ticket (4 minutes)

A STORY OF RATIOS — Lesson 16 — 6•3

Name _____   Date _____

Lesson 16: Symmetry in the Coordinate Plane

Exit Ticket

1. How are the ordered pairs $(4, 9)$ and $(4, -9)$ similar, and how are they different? Are the two points related by a reflection over an axis in the coordinate plane? If so, indicate which axis is the line of symmetry between the points. If they are not related by a reflection over an axis in the coordinate plane, explain how you know.

2. Given the point $(-5, 2)$, write the coordinates of a point that is related by a reflection over the x- or y-axis. Specify which axis is the line of symmetry.

Lesson 16:   Symmetry in the Coordinate Plane

A STORY OF RATIOS — Lesson 16 6•3

Exit Ticket Sample Solutions

1. How are the ordered pairs $(4, 9)$ and $(4, -9)$ similar, and how are they different? Are the two points related by a reflection over an axis in the coordinate plane? If so, indicate which axis is the line of symmetry between the points. If they are not related by a reflection over an axis in the coordinate plane, explain how you know.

   The x-coordinates are the same, but the y-coordinates are opposites, meaning they are the same distance from zero on the x-axis and the same distance but on opposite sides of zero on the y-axis. Reflecting about the x-axis interchanges these two points.

2. Given the point $(-5, 2)$, write the coordinates of a point that is related by a reflection over the x- or y-axis. Specify which axis is the line of symmetry.

   Using the x-axis as a line of symmetry, $(-5, -2)$; using the y-axis as a line of symmetry, $(5, 2)$

Problem Set Sample Solutions

1. Locate a point in Quadrant IV of the coordinate plane. Label the point A, and write its ordered pair next to it.

   Answers will vary; Quadrant IV $(5, -3)$

   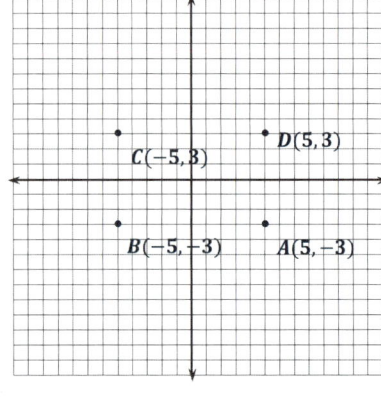

   a. Reflect point A over an axis so that its image is in Quadrant III. Label the image B, and write its ordered pair next to it. Which axis did you reflect over? What is the only difference in the ordered pairs of points A and B?

   $B(-5, -3)$; *reflected over the y-axis*

   The ordered pairs differ only by the sign of their x-coordinates: $A(5, -3)$ and $B(-5, -3)$.

   b. Reflect point B over an axis so that its image is in Quadrant II. Label the image C, and write its ordered pair next to it. Which axis did you reflect over? What is the only difference in the ordered pairs of points B and C? How does the ordered pair of point C relate to the ordered pair of point A?

   $C(-5, 3)$; *reflected over the x-axis*

   The ordered pairs differ only by the signs of their y-coordinates: $B(-5, -3)$ and $C(-5, 3)$.

   The ordered pair for point C differs from the ordered pair for point A by the signs of both coordinates: $A(5, -3)$ and $C(-5, 3)$.

   c. Reflect point C over an axis so that its image is in Quadrant I. Label the image D, and write its ordered pair next to it. Which axis did you reflect over? How does the ordered pair for point D compare to the ordered pair for point C? How does the ordered pair for point D compare to points A and B?

   $D(5, 3)$; *reflected over the y-axis again*

   Point D differs from point C by only the sign of its x-coordinate: $D(5, 3)$ and $C(-5, 3)$.

   Point D differs from point B by the signs of both coordinates: $D(5, 3)$ and $B(-5, -3)$.

   Point D differs from point A by only the sign of the y-coordinate: $D(5, 3)$ and $A(5, -3)$.

A STORY OF RATIOS Lesson 16 6•3

2. Bobbie listened to her teacher's directions and navigated from the point $(-1, 0)$ to $(5, -3)$. She knows that she has the correct answer, but she forgot part of the teacher's directions. Her teacher's directions included the following:

"Move 7 units down, reflect about the __?__-axis, move up 4 units, and then move right 4 units."

Help Bobbie determine the missing axis in the directions, and explain your answer.

The missing line is a reflection over the y-axis. The first line would move the location to $(-1, -7)$. A reflection over the y-axis would move the location to $(1, -7)$ in Quadrant IV, which is 4 units left and 4 units down from the end point $(5, -3)$.

Lesson 16: Symmetry in the Coordinate Plane

A STORY OF RATIOS

Lesson 17 6•3

 Lesson 17: Drawing the Coordinate Plane and Points on the Plane

Student Outcomes

- Students draw a coordinate plane on graph paper in two steps: (1) Draw and label the horizontal and vertical axes; (2) Mark the number scale on each axis.
- Given some points as ordered pairs, students make reasonable choices for scales on both axes and locate and label the points on graph paper.

Classwork

Opening Exercise (5 minutes)

Instruct students to draw all necessary components of the coordinate plane on the blank 20×20 grid provided below, placing the origin at the center of the grid and letting each grid line represent 1 unit. Observe students as they complete the task, using their prior experience with the coordinate plane.

> **Opening Exercise**
>
> Draw all necessary components of the coordinate plane on the blank 20×20 grid provided below, placing the origin at the center of the grid and letting each grid line represent 1 unit.
>
>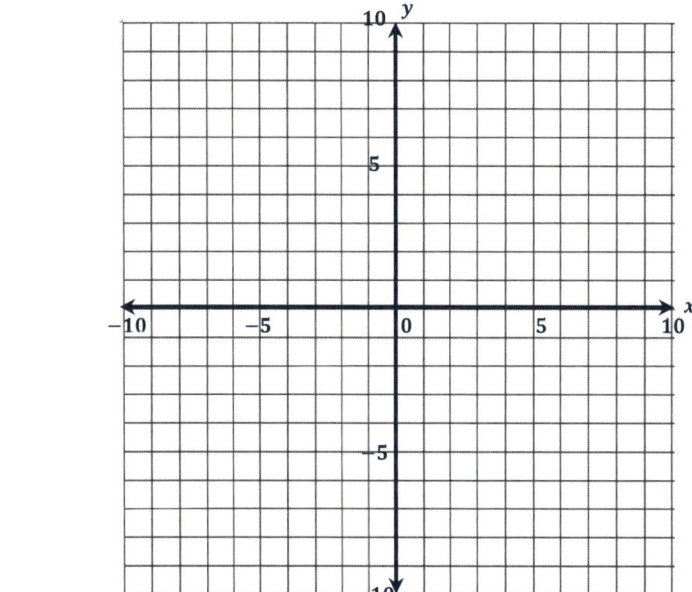

A STORY OF RATIOS   Lesson 17 6•3

MP.4

Students and the teacher together discuss the need for every coordinate plane to have the following:

- The x- and y-axes drawn using a straightedge,
- The horizontal axis labeled x,
- The vertical axis labeled y,
- Each axis labeled using an appropriate scale as dictated by the problem or set of ordered pairs to be graphed.

Students should erase errors and make any necessary changes before proceeding to Example 1.

Example 1 (8 minutes): Drawing the Coordinate Plane Using a 1: 1 Scale

- Is the size of the coordinate grid that we discussed in the Opening Exercise sufficient to graph the points given in the set in Example 1?
  □ Yes. All x- and y-coordinates are between -10 and 10, and both axes on the grid range from -10 to 10.

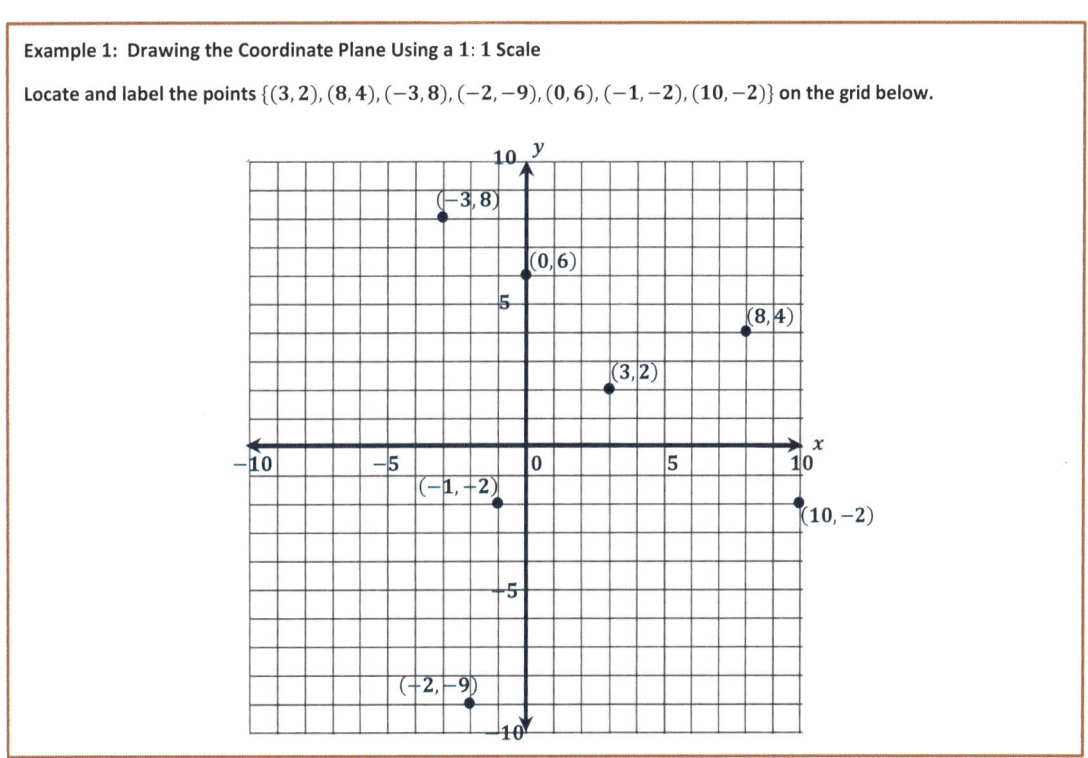

Example 1: Drawing the Coordinate Plane Using a 1: 1 Scale

Locate and label the points $\{(3,2), (8,4), (-3,8), (-2,-9), (0,6), (-1,-2), (10,-2)\}$ on the grid below.

- Can you name a point that could not be located on this grid? Explain.
  □ The point $(18, 5)$ could not be located on this grid because 18 is greater than 10 and, therefore, to the right of 10 on the x-axis. 10 is the greatest number shown on this grid.
- Discuss ways in which the point $(18, 5)$ could be graphed without changing the size of the grid.
  □ Changing the number of units that each grid line represents would allow us to fit greater numbers on the axes. Changing the number of units per grid line to 2 units would allow a range of -20 to 20 on the x-axis.

Lesson 17: Drawing the Coordinate Plane and Points on the Plane

A STORY OF RATIOS — Lesson 17 — 6•3

Example 2 (8 minutes): Drawing the Coordinate Plane Using an Increased Number Scale for One Axis

Students increase the number of units represented by each grid line in the coordinate plane in order to graph a given set of ordered pairs.

- Examine the given points. What is the range of values used as x-coordinates? How many units should we assign per grid line to show this range of values? Explain.
    - *The x-coordinates range from -4 to 9, all within the range of -10 to 10, so we will assign each grid line to represent 1 unit.*
- What is the range of values used as y-coordinates? How many units should we assign per grid line to show this range of values? Explain.
    - *The y-coordinates range from -40 to 35. If we let each grid line represent 5 units, then the y-axis will include the range -50 to 50.*
- Draw and label the coordinate plane, and then locate and label the set of points.

> **Example 2:** Drawing the Coordinate Plane Using an Increased Number Scale for One Axis
>
> Draw a coordinate plane on the grid below, and then locate and label the following points:
>
> $$\{(-4, 20), (-3, 35), (1, -35), (6, 10), (9, -40)\}.$$
>
>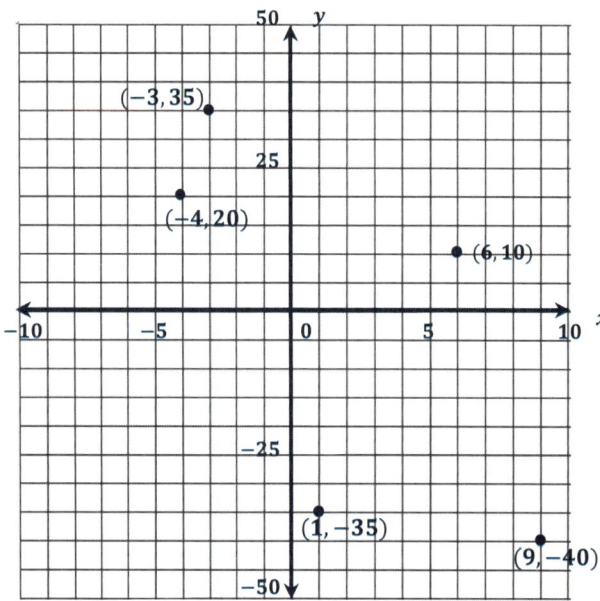

Lesson 17: Drawing the Coordinate Plane and Points on the Plane

A STORY OF RATIOS Lesson 17 6•3

Example 3 (8 minutes): Drawing the Coordinate Plane Using a Decreased Number Scale for One Axis

Students divide units among multiple grid lines in the coordinate plane in order to graph a given set of ordered pairs.

- Examine the given points. Will either the x- or y-coordinates require a change of scale in the plane? Explain.
    - *The x-coordinates range from -0.7 to 0.8. This means that if each grid line represents one unit, the points would all be very close to the y-axis, making it difficult to interpret.*
- How could we change the number of units represented per grid line to better show the points in the given set?
    - *Divide 1 unit into tenths so that each grid line represents a tenth of a unit, and the x-axis then ranges from -1 to 1.*
- Draw and label the coordinate plane, and then locate and label the set of points.

Example 3: Drawing the Coordinate Plane Using a Decreased Number Scale for One Axis

Draw a coordinate plane on the grid below, and then locate and label the following points:

$$\{(0.1, 4), (0.5, 7), (-0.7, -5), (-0.4, 3), (0.8, 1)\}.$$

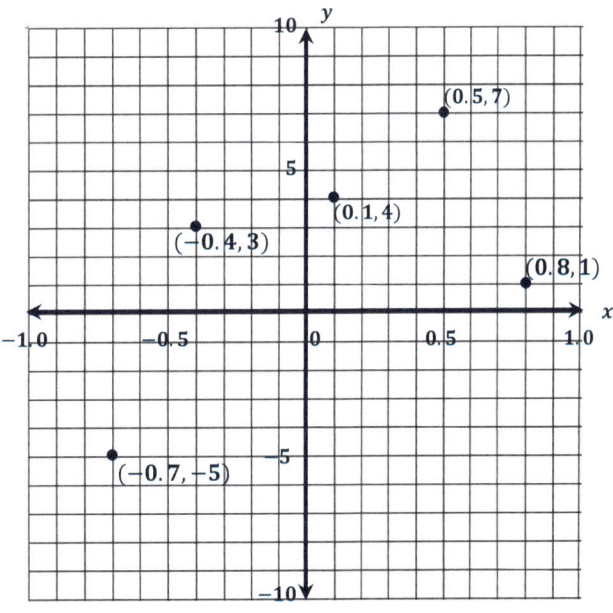

Lesson 17: Drawing the Coordinate Plane and Points on the Plane

A STORY OF RATIOS

Lesson 17   6•3

Example 4 (8 minutes): Drawing the Coordinate Plane Using a Different Number Scale for Both Axes

Students appropriately scale the axes in the coordinate plane in order to graph a given set of ordered pairs. Note that the provided grid is 16 × 16 with fewer grid lines than the previous examples.

> **Example 4: Drawing the Coordinate Plane Using a Different Number Scale for Both Axes**
>
> Determine a scale for the x-axis that will allow all x-coordinates to be shown on your grid.
>
> The grid is 16 units wide, and the x-coordinates range from -14 to 14. If I let each grid line represent 2 units, then the x-axis will range from -16 to 16.
>
> Determine a scale for the y-axis that will allow all y-coordinates to be shown on your grid.
>
> The grid is 16 units high, and the y-coordinates range from -4 to 3.5. I could let each grid line represent one unit, but if I let each grid line represent $\frac{1}{2}$ of a unit, the points will be easier to graph.
>
> Draw and label the coordinate plane, and then locate and label the set of points.
>
> $$\{(-14, 2), (-4, -0.5), (6, -3.5), (14, 2.5), (0, 3.5), (-8, -4)\}$$
>
>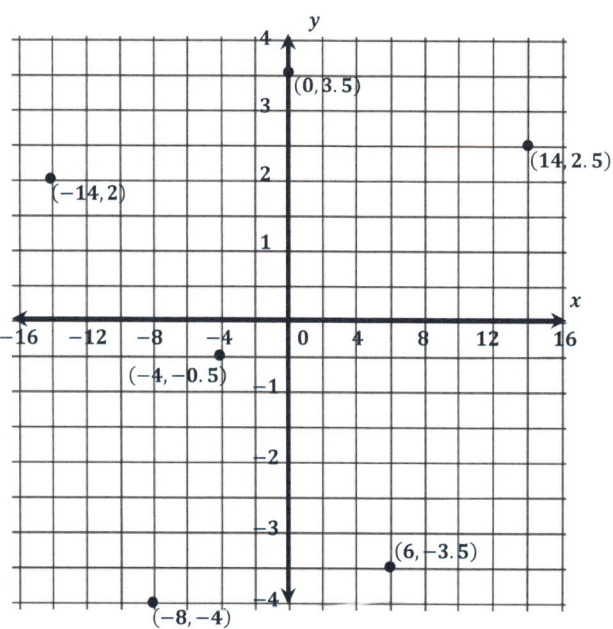

- How was this example different from the first three examples in this lesson?
  - *The given set of points caused me to change the scales on both axes, and the given grid had fewer grid lines.*
- Did these differences affect your decision making as you created the coordinate plane? Explain.
  - *Shrinking the scale of the x-axis allowed me to show a larger range of numbers, but fewer grid lines limited that range.*

Closing (4 minutes)

- Why is it important to label the axes when setting up a coordinate plane?
  - *It is important to label the axes when setting up a coordinate plane so that the person viewing the graph knows which axis represents which coordinate and what scale is being used. If a person does not know the scale being used, she will likely misinterpret the graph.*
- Why shouldn't you draw and label the entire coordinate grid before looking at the points to be graphed?
  - *Looking at the range of values in a given set of points allows you to determine an appropriate scale. If you set a scale before observing the given values, you will likely have to change the scale on your axes.*

Lesson Summary

- The axes of the coordinate plane must be drawn using a straightedge and labeled x (horizontal axis) and y (vertical axis).
- Before assigning a scale to the axes, it is important to assess the range of values found in a set of points as well as the number of grid lines available. This allows you to determine an appropriate scale so all points can be represented on the coordinate plane that you construct.

Exit Ticket (4 minutes)

A STORY OF RATIOS   Lesson 17   6•3

Name _____ Date _____

Lesson 17: Drawing the Coordinate Plane and Points on the Plane

Exit Ticket

Determine an appropriate scale for the set of points given below. Draw and label the coordinate plane, and then locate and label the set of points.

$$\{(10, 0.2), (-25, 0.8), (0, -0.4), (20, 1), (-5, -0.8)\}$$

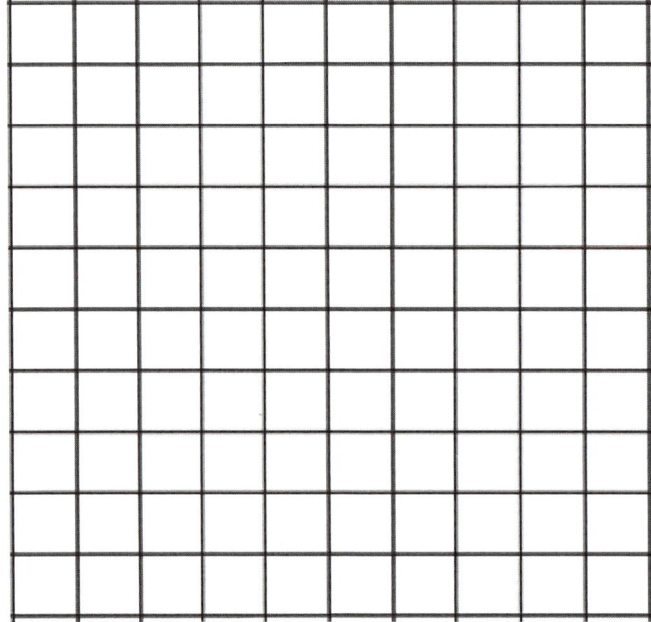

Exit Ticket Sample Solutions

Determine an appropriate scale for the set of points given below. Draw and label the coordinate plane, and then locate and label the set of points.

$$\{(10, 0.2), (-25, 0.8), (0, -0.4), (20, 1), (-5, -0.8)\}$$

The x-coordinates range from -25 to 20. The grid is 10 units wide. If I let each grid line represent 5 units, then the x-axis will range from -25 to 25.

The y-coordinates range from -0.8 to 1. The grid is 10 units high. If I let each grid line represent two-tenths of a unit, then the y-axis will range from -1 to 1.

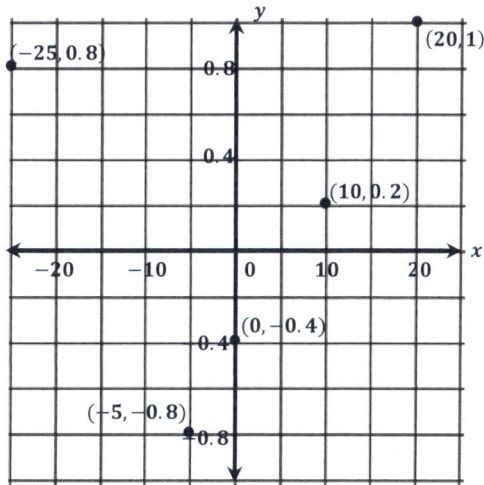

Problem Set Sample Solutions

1. Label the coordinate plane, and then locate and label the set of points below.

$$\{(0.3, 0.9), (-0.1, 0.7), (-0.5, -0.1), (-0.9, 0.3), (0, -0.4)\}$$

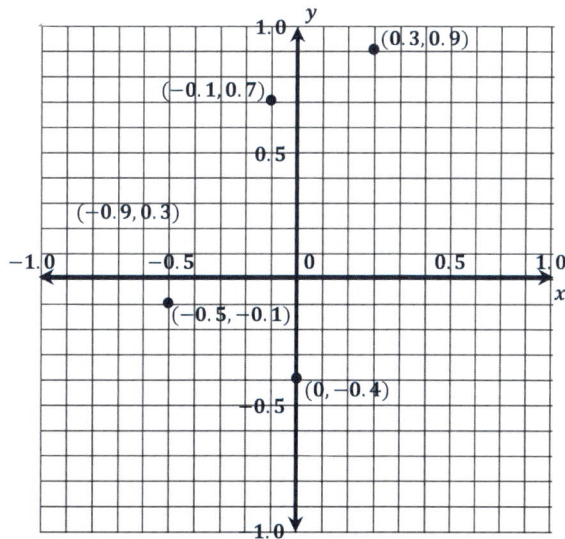

Lesson 17: Drawing the Coordinate Plane and Points on the Plane

2. Label the coordinate plane, and then locate and label the set of points below.

$$\{(90, 9), (-110, -11), (40, 4), (-60, -6), (-80, -8)\}$$

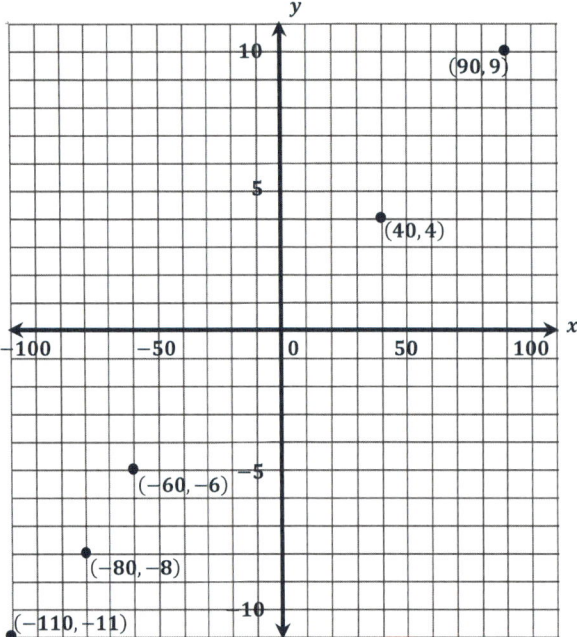

Extension:

3. Describe the pattern you see in the coordinates in Problem 2 and the pattern you see in the points. Are these patterns consistent for other points too?

   The x-coordinate for each of the given points is 10 times its y-coordinate. When I graphed the points, they appear to make a straight line. I checked other ordered pairs with the same pattern, such as $(-100, -10)$, $(20, 2)$, and even $(0, 0)$, and it appears that these points are also on that line.

A STORY OF RATIOS   Lesson 18   6•3

 Lesson 18: Distance on the Coordinate Plane

Student Outcomes

- Students compute the length of horizontal and vertical line segments with integer coordinates for end points in the coordinate plane by counting the number of units between end points and using absolute value.

Classwork

Opening Exercise (5 minutes)

> **Opening Exercise**
>
> Four friends are touring on motorcycles. They come to an intersection of two roads; the road they are on continues straight, and the other is perpendicular to it. The sign at the intersection shows the distances to several towns. Draw a map/diagram of the roads, and use it and the information on the sign to answer the following questions:
>
> Albertsville ← 8 mi.
> Blossville ↑ 3 mi.
> Cheyenne ↑ 12 mi.
> Dewey Falls → 6 mi.
>
> What is the distance between Albertsville and Dewey Falls?
>
> *Students draw and use their maps to answer. Albertsville is 8 miles to the left, and Dewey Falls is 6 miles to the right. Since the towns are in opposite directions from the intersection, their distances must be combined by addition, $8 + 6 = 14$, so the distance between Albertsville and Dewey Falls is 14 miles.*
>
> What is the distance between Blossville and Cheyenne?
>
> *Blossville and Cheyenne are both straight ahead from the intersection in the direction that they are going. Since they are on the same side of the intersection, Blossville is on the way to Cheyenne, so the distance to Cheyenne includes the 3 miles to Blossville. To find the distance from Blossville to Cheyenne, I have to subtract; $12 - 3 = 9$. So, the distance from Blossville to Cheyenne is 9 miles.*
>
> On the coordinate plane, what represents the intersection of the two roads?
>
> *The intersection is represented by the origin.*

Example 1 (6 minutes): The Distance Between Points on an Axis

Students find the distance between points on the x-axis by finding the distance between numbers on the number line. They find the absolute values of the x-coordinates and add or subtract their absolute values to determine the distance between the points.

> **Example 1: The Distance Between Points on an Axis**
>
> Consider the points $(-4, 0)$ and $(5, 0)$.
>
> What do the ordered pairs have in common, and what does that mean about their location in the coordinate plane?
>
> *Both of their y-coordinates are zero, so each point lies on the x-axis, the horizontal number line.*

   Lesson 18:   Distance on the Coordinate Plane   175

A STORY OF RATIOS   Lesson 18   6•3

> How did we find the distance between two numbers on the number line?
>
> *We calculated the absolute values of the numbers, which told us how far the numbers were from zero. If the numbers were located on opposite sides of zero, then we added their absolute values together. If the numbers were located on the same side of zero, then we subtracted their absolute values.*
>
> Use the same method to find the distance between $(-4, 0)$ and $(5, 0)$.
>
> $|-4| = 4$ *and* $|5| = 5$. *The numbers are on opposite sides of zero, so the absolute values get combined:* $4 + 5 = 9$. *The distance between* $(-4, 0)$ *and* $(5, 0)$ *is 9 units.*

Example 2 (5 minutes): The Length of a Line Segment on an Axis

Students find the length of a line segment that lies on the y-axis by finding the distance between its end points.

> **Example 2: The Length of a Line Segment on an Axis**
>
> Consider the line segment with end points $(0, -6)$ and $(0, -11)$.
>
> What do the ordered pairs of the end points have in common, and what does that mean about the line segment's location in the coordinate plane?
>
> *The x-coordinates of both end points are zero, so the points lie on the y-axis, the vertical number line. If its end points lie on a vertical number line, then the line segment itself must also lie on the vertical line.*
>
> Find the length of the line segment described by finding the distance between its end points $(0, -6)$ and $(0, -11)$.
>
> $|-6| = 6$ *and* $|-11| = 11$. *The numbers are on the same side of zero, which means the longer distance contains the shorter distance, so the absolute values need to be subtracted:* $11 - 6 = 5$. *The distance between* $(0, -6)$ *and* $(0, -11)$ *is 5 units, so the length of the line segment with end points* $(0, -6)$ *and* $(0, -11)$ *is 5 units.*

Example 3 (10 minutes): Length of a Horizontal or Vertical Line Segment That Does Not Lie on an Axis

Students find the length of a vertical line segment that does not lie on the y-axis by finding the distance between its end points.

MP.7

> **Example 3: Length of a Horizontal or Vertical Line Segment That Does Not Lie on an Axis**
>
> Consider the line segment with end points $(-3, 3)$ and $(-3, -5)$.
>
> What do the end points, which are represented by the ordered pairs, have in common? What does that tell us about the location of the line segment on the coordinate plane?
>
> *Both end points have x-coordinates of -3, so the points lie on the vertical line that intersects the x-axis at -3. This means that the end points of the line segment, and thus the line segment, lie on a vertical line.*
>
> Find the length of the line segment by finding the distance between its end points.
>
> *The end points are on the same vertical line, so we only need to find the distance between 3 and -5 on the number line. $|3| = 3$ and $|-5| = 5$, and the numbers are on opposite sides of zero, so the values must be added: $3 + 5 = 8$. So, the distance between $(-3, 3)$ and $(-3, -5)$ is 8 units.*

Scaffolding:

Students may need to draw an auxiliary line through the end points to help visualize a horizontal or vertical number line.

Exercise (10 minutes)

Students calculate the distance between pairs of points using absolute values.

176   Lesson 18:   Distance on the Coordinate Plane

This work is derived from Eureka Math ™ and licensed by Great Minds. ©2015 Great Minds. eureka-math.org
G6-M3-TE-B3-1.3.0-07.2015

A STORY OF RATIOS   Lesson 18   6•3

> **Exercise**
>
> Find the lengths of the line segments whose end points are given below. Explain how you determined that the line segments are horizontal or vertical.
>
> a. $(-3, 4)$ and $(-3, 9)$
>
>   *Both end points have x-coordinates of -3, so the points lie on a vertical line that passes through -3 on the x-axis. $|4| = 4$ and $|9| = 9$, and the numbers are on the same side of zero. By subtraction, $9 - 4 = 5$, so the length of the line segment with end points $(-3, 4)$ and $(-3, 9)$ is 5 units.*
>
> b. $(2, -2)$ and $(-8, -2)$
>
>   *Both end points have y-coordinates of -2, so the points lie on a horizontal line that passes through -2 on the y-axis. $|2| = 2$ and $|-8| = 8$, and the numbers are on opposite sides of zero, so the absolute values must be added. By addition, $8 + 2 = 10$, so the length of the line segment with end points $(2, -2)$ and $(-8, -2)$ is 10 units.*
>
> c. $(-6, -6)$ and $(-6, 1)$
>
>   *Both end points have x-coordinates of -6, so the points lie on a vertical line. $|-6| = 6$ and $|1| = 1$, and the numbers are on opposite sides of zero, so the absolute values must be added. By addition, $6 + 1 = 7$, so the length of the line segment with end points $(-6, -6)$ and $(-6, 1)$ is 7 units.*
>
> d. $(-9, 4)$ and $(-4, 4)$
>
>   *Both end points have y-coordinates of 4, so the points lie on a horizontal line. $|-9| = 9$ and $|-4| = 4$, and the numbers are on the same side of zero. By subtraction, $9 - 4 = 5$, so the length of the line segment with end points $(-9, 4)$ and $(-4, 4)$ is 5 units.*
>
> e. $(0, -11)$ and $(0, 8)$
>
>   *Both end points have x-coordinates of 0, so the points lie on the y-axis. $|-11| = 11$ and $|8| = 8$, and the numbers are on opposite sides of zero, so their absolute values must be added. By addition, $11 + 8 = 19$, so the length of the line segment with end points $(0, -11)$ and $(0, 8)$ is 19 units.*

Closing (3 minutes)

- Why is it possible for us to find the length of a horizontal or vertical line segment even if it's not on the x- or y-axis?
    - *A line can still be a horizontal or vertical line even if it is not on the x- or y-axis; therefore, we can still use the same strategy.*
- Can you think of a real-world situation where this might be useful?
    - *Finding the distance on a map*

> **Lesson Summary**
>
> To find the distance between points that lie on the same horizontal line or on the same vertical line, we can use the same strategy that we used to find the distance between points on the number line.

Exit Ticket (6 minutes)

Lesson 18:   Distance on the Coordinate Plane   177

Name _____   Date _____

Lesson 18: Distance on the Coordinate Plane

Exit Ticket

Determine whether each given pair of end points lies on the same horizontal or vertical line. If so, find the length of the line segment that joins the pair of points. If not, explain how you know the points are not on the same horizontal or vertical line.

a. $(0, -2)$ and $(0, 9)$

b. $(11, 4)$ and $(2, 11)$

c. $(3, -8)$ and $(3, -1)$

d. $(-4, -4)$ and $(5, -4)$

A STORY OF RATIOS　　　　　　　　　　　　　　　　　　　　　　　　　Lesson 18　6•3

Exit Ticket Sample Solutions

Determine whether each given pair of end points lies on the same horizontal or vertical line. If so, find the length of the line segment that joins the pair of points. If not, explain how you know the points are not on the same horizontal or vertical line.

a. $(0, -2)$ and $(0, 9)$

The end points both have x-coordinates of 0, so they both lie on the y-axis, which is a vertical line. They lie on opposite sides of zero, so their absolute values have to be combined to get the total distance. $|-2| = 2$ and $|9| = 9$, so by addition, $2 + 9 = 11$. The length of the line segment with end points $(0, -2)$ and $(0, 9)$ is 11 units.

b. $(11, 4)$ and $(2, 11)$

The points do not lie on the same horizontal or vertical line because they do not share a common x- or y-coordinate.

c. $(3, -8)$ and $(3, -1)$

The end points both have x-coordinates of 3, so the points lie on a vertical line that passes through 3 on the x-axis. The y-coordinates lie on the same side of zero. The distance between the points is determined by subtracting their absolute values, $|-8| = 8$ and $|-1| = 1$. So, by subtraction, $8 - 1 = 7$. The length of the line segment with end points $(3, -8)$ and $(3, -1)$ is 7 units.

d. $(-4, -4)$ and $(5, -4)$

The end points have the same y-coordinate of -4, so they lie on a horizontal line that passes through -4 on the y-axis. The numbers lie on opposite sides of zero on the number line, so their absolute values must be added to obtain the total distance, $|-4| = 4$ and $|5| = 5$. So, by addition, $4 + 5 = 9$. The length of the line segment with end points $(-4, -4)$ and $(5, -4)$ is 9 units.

Problem Set Sample Solutions

1. Find the length of the line segment with end points $(7, 2)$ and $(-4, 2)$, and explain how you arrived at your solution.

   11 units. Both points have the same y-coordinate, so I knew they were on the same horizontal line. I found the distance between the x-coordinates by counting the number of units on a horizontal number line from -4 to zero and then from zero to 7, and $4 + 7 = 11$.

   or

   I found the distance between the x-coordinates by finding the absolute value of each coordinate. $|7| = 7$ and $|-4| = 4$. The coordinates lie on opposite sides of zero, so I found the length by adding the absolute values together. Therefore, the length of a line segment with end points $(7, 2)$ and $(-4, 2)$ is 11 units.

2. Sarah and Jamal were learning partners in math class and were working independently. They each started at the point $(-2, 5)$ and moved 3 units vertically in the plane. Each student arrived at a different end point. How is this possible? Explain and list the two different end points.

   It is possible because Sarah could have counted up and Jamal could have counted down or vice versa. Moving 3 units in either direction vertically would generate the following possible end points: $(-2, 8)$ or $(-2, 2)$.

3. The length of a line segment is 13 units. One end point of the line segment is $(-3, 7)$. Find four points that could be the other end points of the line segment.

   $(-3, 20), (-3, -6), (-16, 7)$ or $(10, 7)$

Lesson 18: Distance on the Coordinate Plane

A STORY OF RATIOS   Lesson 19   6•3

 Lesson 19: Problem Solving and the Coordinate Plane

Student Outcomes

- Students solve problems related to the distance between points that lie on the same horizontal or vertical line.
- Students use the coordinate plane to graph points, line segments, and geometric shapes in the various quadrants and then use the absolute value to find the related distances.

Lesson Notes

The grid provided in the Opening Exercise is also used for Exercises 1–6 since each exercise is sequential. Students extend their knowledge about finding distances between points on the coordinate plane to the associated lengths of line segments and sides of geometric figures.

Classwork

Opening Exercise (3 minutes)

> **Opening Exercise**
>
> In the coordinate plane, find the distance between the points using absolute value.
>
> *The distance between the points is 8 units. The points have the same first coordinates and, therefore, lie on the same vertical line. $|-3| = 3$, and $|5| = 5$, and the numbers lie on opposite sides of 0, so their absolute values are added together; $3 + 5 = 8$. We can check our answer by just counting the number of units between the two points.*
>
>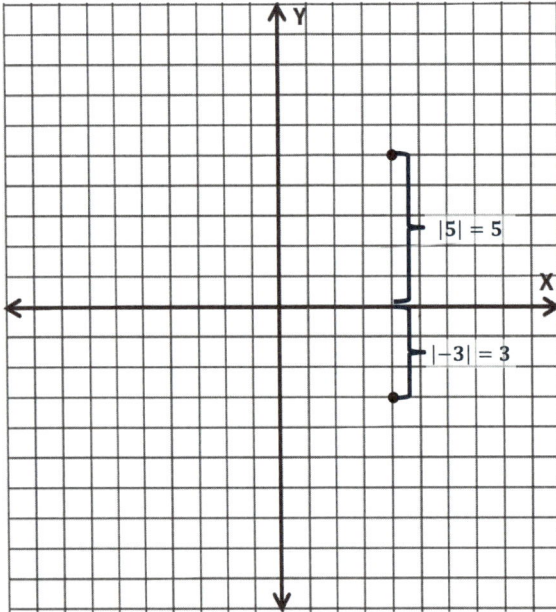

180   Lesson 19:   Problem Solving and the Coordinate Plane

A STORY OF RATIOS   Lesson 19   6•3

Exploratory Challenge

Exercises 1–2 (8 minutes): The Length of a Line Segment Is the Distance Between Its End Points

Students relate the distance between two points lying in different quadrants of the coordinate plane to the length of a line segment with those end points. Students then use this relationship to graph a horizontal or vertical line segment using distance to find the coordinates of end points.

Exploratory Challenge

Exercises 1–2: The Length of a Line Segment Is the Distance Between Its End Points

1. Locate and label $(4, 5)$ and $(4, -3)$. Draw the line segment between the end points given on the coordinate plane. How long is the line segment that you drew? Explain.

   The length of the line segment is also 8 units. I found that the distance between $(4, -3)$ and $(4, 5)$ is 8 units. Because the end points are on opposite sides of zero, I added the absolute values of the second coordinates together, so the distance from end to end is 8 units.

2. Draw a horizontal line segment starting at $(4, -3)$ that has a length of 9 units. What are the possible coordinates of the other end point of the line segment? (There is more than one answer.)

   $(-5, -3)$ or $(13, -3)$

   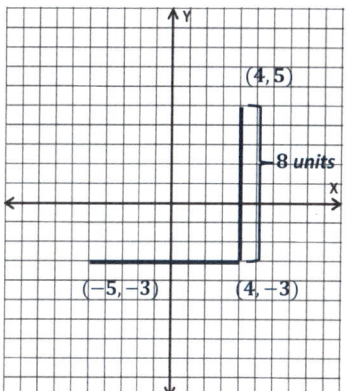

   Which point did you choose to be the other end point of the horizontal line segment? Explain how and why you chose that point. Locate and label the point on the coordinate grid.

   The other end point of the horizontal line segment is $(-5, -3)$. I chose this point because the other option, $(13, -3)$, is located off of the given coordinate grid.
   Note: Students may choose the end point $(13, -3)$, but they must change the number scale of the x-axis to do so.

Exercise 3 (5 minutes): Extending Lengths of Line Segments to Sides of Geometric Figures

Exercise 3: Extending Lengths of Line Segments to Sides of Geometric Figures

3. The two line segments that you have just drawn could be seen as two sides of a rectangle. Given this, the end points of the two line segments would be three of the vertices of this rectangle.

   a. Find the coordinates of the fourth vertex of the rectangle. Explain how you find the coordinates of the fourth vertex using absolute value.

      The fourth vertex is $(-5, 5)$. The opposite sides of a rectangle are the same length, so the length of the vertical side starting at $(-5, -3)$ has to be 8 units long. Also, the side from $(-5, -3)$ to the remaining vertex is a vertical line, so the end points must have the same first coordinate. $|-3| = 3$, and $8 - 3 = 5$, so the remaining vertex must be five units above the x-axis.
      Note: Students can use a similar argument using the length of the horizontal side starting at $(4, 5)$, knowing it has to be 9 units long.

   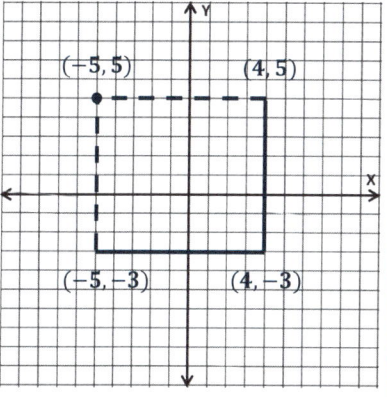

Lesson 19:   Problem Solving and the Coordinate Plane   181

A STORY OF RATIOS

Lesson 19 6•3

b. How does the fourth vertex that you found relate to each of the consecutive vertices in either direction? Explain.

The fourth vertex has the same first coordinate as $(-5, -3)$ because they are the end points of a vertical line segment. The fourth vertex has the same second coordinate as $(4, 5)$ since they are the end points of a horizontal line segment.

c. Draw the remaining sides of the rectangle.

Exercises 4–6 (6 minutes): Using Lengths of Sides of Geometric Figures to Solve Problems

Exercises 4–6: Using Lengths of Sides of Geometric Figures to Solve Problems

4. Using the vertices that you have found and the lengths of the line segments between them, find the perimeter of the rectangle.

   $8 + 9 + 8 + 9 = 34$; *the perimeter of the rectangle is 34 units.*

Scaffolding:

Students may need to review and discuss the concepts of perimeter and area from earlier grades.

5. Find the area of the rectangle.

   $9 \times 8 = 72$; *the area of the rectangle is 72 units².*

6. Draw a diagonal line segment through the rectangle with opposite vertices for end points. What geometric figures are formed by this line segment? What are the areas of each of these figures? Explain.

   The diagonal line segment cuts the rectangle into two right triangles. The areas of the triangles are 36 units² each because the triangles each make up half of the rectangle, and half of 72 is 36.

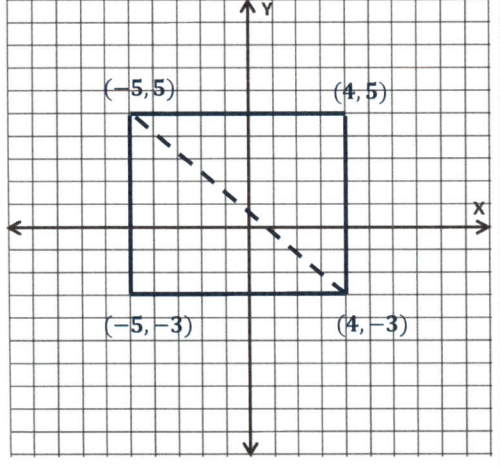

Extension (If time allows): Line the edge of a piece of paper up to the diagonal in the rectangle. Mark the length of the diagonal on the edge of the paper. Align your marks horizontally or vertically on the grid, and estimate the length of the diagonal to the nearest integer. Use that estimation to now estimate the perimeter of the triangles.

The length of the diagonal is approximately 12 units, and the perimeter of each triangle is approximately 29 units.

Exercise 7 (8 minutes)

MP.1

Exercise 7

7. Construct a rectangle on the coordinate plane that satisfies each of the criteria listed below. Identify the coordinate of each of its vertices.

   - Each of the vertices lies in a different quadrant.
   - Its sides are either vertical or horizontal.
   - The perimeter of the rectangle is 28 units.

Answers will vary. The example to the right shows a rectangle with side lengths 10 and 4 units. The coordinates of the rectangle's vertices are $(-6, 3)$, $(4, 3)$, $(4, -1)$, and $(-6, -1)$.

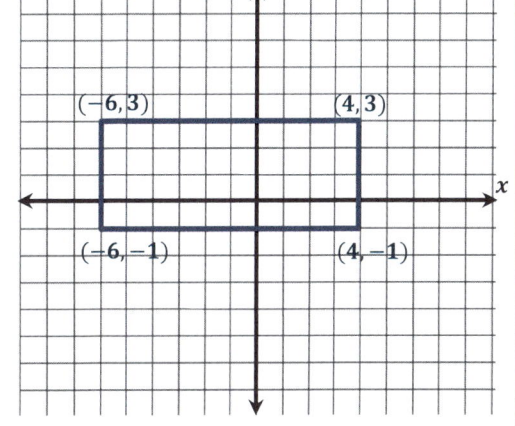

Using absolute value, show how the lengths of the sides of your rectangle provide a perimeter of 28 units.

$|-6| = 6$, $|4| = 4$, and $6 + 4 = 10$, so the width of my rectangle is 10 units.

$|3| = 3$, $|-1| = 1$, and $3 + 1 = 4$, so the height of my rectangle is 4 units.

$10 + 4 + 10 + 4 = 28$, so the perimeter of my rectangle is 28 units.

Closing (5 minutes)

- How do we determine the length of a horizontal line segment whose end points lie in different quadrants of the coordinate plane?
  - *If the points are in different quadrants, then the x-coordinates lie on opposite sides of zero. The distance between the x-coordinates can be found by adding the absolute values of the x-coordinates. (The y-coordinates are the same and show that the points lie on a horizontal line.)*

- If we know one end point of a vertical line segment and the length of the line segment, how do we find the other end point of the line segment? Is the process the same with a horizontal line segment?
  - *If the line segment is vertical, then the other end point could be above or below the given end point. If we know the length of the line segment, then we can count up or down from the given end point to find the other end point. We can check our answer using the absolute values of the y-coordinates. The process is similar with a horizontal line. If we know the length of the line segment, then we can count to the left or the right from the given end point to find the other end point.*

Lesson Summary

- The length of a line segment on the coordinate plane can be determined by finding the distance between its end points.
- You can find the perimeter and area of figures such as rectangles and right triangles by finding the lengths of the line segments that make up their sides and then using the appropriate formula.

Exit Ticket (10 minutes)

Lesson 19: Problem Solving and the Coordinate Plane

Name _____ Date _____

Lesson 19: Problem Solving and the Coordinate Plane

Exit Ticket

1. The coordinates of one end point of a line segment are $(-2, -7)$. The line segment is 12 units long. Give three possible coordinates of the line segment's other end point.

2. Graph a rectangle with an area of 12 units^2 such that its vertices lie in at least two of the four quadrants in the coordinate plane. State the lengths of each of the sides, and use absolute value to show how you determined the lengths of the sides.

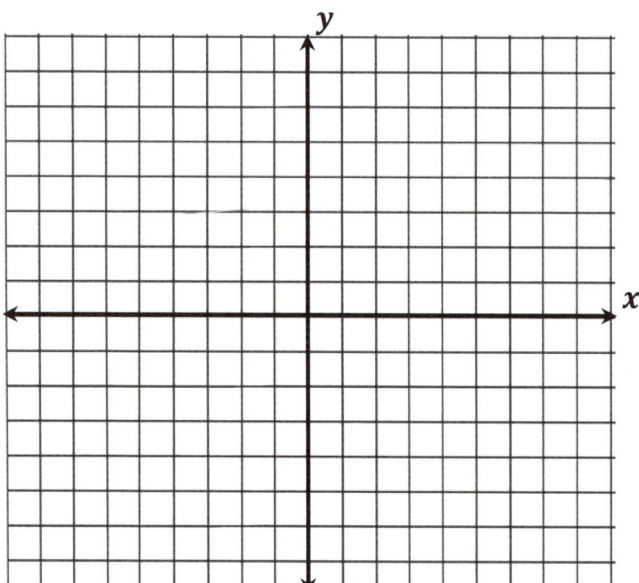

Exit Ticket Sample Solutions

1. The coordinates of one end point of a line segment are $(-2, -7)$. The line segment is 12 units long. Give three possible coordinates of the line segment's other end point.

   $(10, -7); (-14, -7); (-2, 5); (-2, -19)$

2. Graph a rectangle with an area of 12 units² such that its vertices lie in at least two of the four quadrants in the coordinate plane. State the lengths of each of the sides, and use absolute value to show how you determined the lengths of the sides.

   Answers will vary. The rectangle can have side lengths of 6 and 2 or 3 and 4. A sample is provided on the grid on the right. $6 \times 2 = 12$

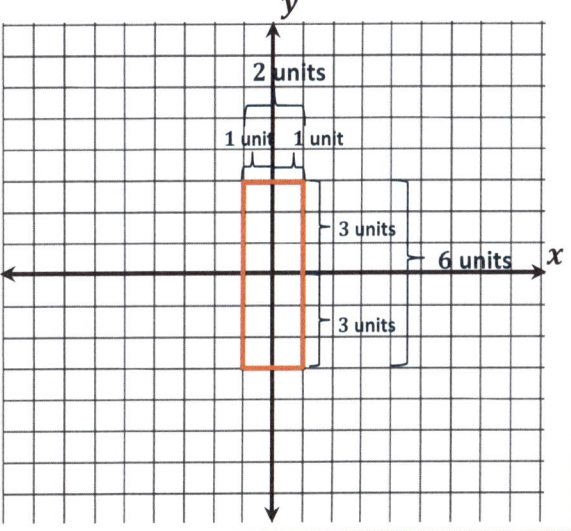

Problem Set Sample Solutions

Please provide students with three coordinate grids to use in completing the Problem Set.

1. One end point of a line segment is $(-3, -6)$. The length of the line segment is 7 units. Find four points that could serve as the other end point of the given line segment.

   $(-10, -6); (4, -6); (-3, 1); (-3, -13)$

2. Two of the vertices of a rectangle are $(1, -6)$ and $(-8, -6)$. If the rectangle has a perimeter of 26 units, what are the coordinates of its other two vertices?

   $(1, -2)$ and $(-8, -2)$, or $(1, -10)$ and $(-8, -10)$

3. A rectangle has a perimeter of 28 units, an area of 48 square units, and sides that are either horizontal or vertical. If one vertex is the point $(-5, -7)$ and the origin is in the interior of the rectangle, find the vertex of the rectangle that is opposite $(-5, -7)$.

   $(1, 1)$

Lesson 19: Problem Solving and the Coordinate Plane

A STORY OF RATIOS End-of-Module Assessment Task 6•3

Name _____   Date _____

1. Mr. Kindle invested some money in the stock market. He tracks his gains and losses using a computer program. Mr. Kindle receives a daily email that updates him on all his transactions from the previous day. This morning, his email read as follows:

   Good morning, Mr. Kindle,

   Yesterday's investment activity included a loss of $800, a gain of $960, and another gain of $230. Log in now to see your current balance.

   a. Write an integer to represent each gain and loss.

Description	Integer Representation
Loss of $800	
Gain of $960	
Gain of $230	

   b. Mr. Kindle noticed that an error had been made on his account. The "loss of $800" should have been a "gain of $800." Locate and label both points that represent "a loss of $800" and "a gain of $800" on the number line below. Describe the relationship of these two numbers when zero represents no change (gain or loss).

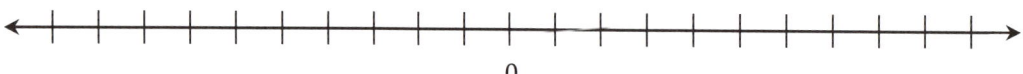
0

c. Mr. Kindle wanted to correct the error, so he entered −(−$800) into the program. He made a note that read, "The opposite of the opposite of $800 is $800." Is his reasoning correct? Explain.

2. At 6:00 a.m., Buffalo, NY, had a temperature of 10°F. At noon, the temperature was −10°F, and at midnight, it was −20°F.

   a. Write a statement comparing −10°F and −20°F.

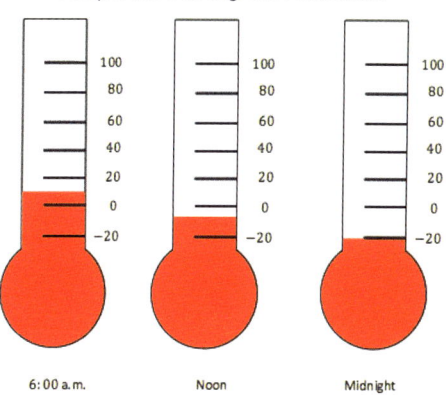

   b. Write an inequality statement that shows the relationship between the three recorded temperatures. Which temperature is the warmest?

c. Explain how to use absolute value to find the number of degrees below zero the temperature was at noon.

d. In Peekskill, NY, the temperature at 6:00 a.m. was −12°F. At noon, the temperature was the exact opposite of Buffalo's temperature at 6:00 a.m. At midnight, a meteorologist recorded the temperature as −6°F in Peekskill. He concluded that "For temperatures below zero, as the temperature increases, the absolute value of the temperature decreases." Is his conclusion valid? Explain and use a vertical number line to support your answer.

3. Choose an integer between 0 and −5 on a number line, and label the point P. Locate and label each of the following points and their values on the number line.

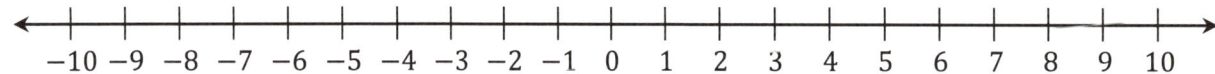

a. Label point A: the opposite of point P.

b. Label point B: a number less than point P.

c. Label point C: a number greater than point P.

d. Label point D: a number halfway between point P and the integer to the right of point P.

End-of-Module Assessment Task 6•3

4. Julia is learning about elevation in math class. She decided to research some facts about New York State to better understand the concept. Here are some facts that she found.

   - *Mount Marcy is the highest point in New York State. It is 5,343 feet above sea level.*
   - *Lake Erie is 210 feet below sea level.*
   - *The elevation of Niagara Falls, NY, is 614 feet above sea level.*
   - *The lobby of the Empire State Building is 50 feet above sea level.*
   - *New York State borders the Atlantic Coast, which is at sea level.*
   - *The lowest point of Cayuga Lake is 435 feet below sea level.*

   a. Write an integer that represents each location in relationship to sea level.

   Mount Marcy   _____

   Lake Erie   _____

   Niagara Falls, NY   _____

   Empire State Building _____

   Atlantic Coast   _____

   Cayuga Lake   _____

   b. Explain what negative and positive numbers tell Julia about elevation.

c. Order the elevations from least to greatest, and then state their absolute values. Use the chart below to record your work.

Elevations	Absolute Values of Elevations

d. Circle the row in the table that represents sea level. Describe how the order of the elevations below sea level compares to the order of their absolute values. Describe how the order of the elevations above sea level compares to the order of their absolute values.

5. For centuries, a mysterious sea serpent has been rumored to live at the bottom of Mysterious Lake. A team of historians used a computer program to plot the last five positions of the sightings.

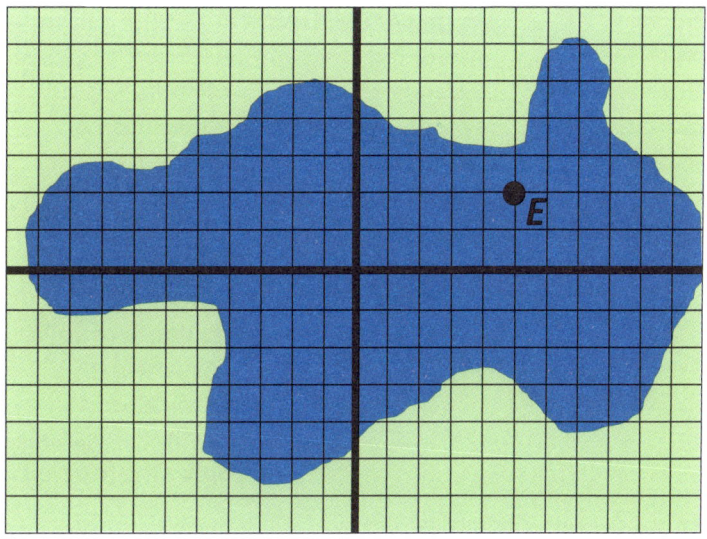

a. Locate and label the locations of the last four sightings: $A\left(-9\frac{1}{2}, 0\right)$, $B(-3, -4.75)$, $C(9, 2)$, and $D(8, -2.5)$.

b. Over time, most of the sightings occurred in Quadrant III. Write the coordinates of a point that lies in Quadrant III.

c. What is the distance between point A and the point $\left(9\frac{1}{2}, 0\right)$? Show your work to support your answer.

d. What are the coordinates of point E on the coordinate plane?

e. Point F is related to point E. Its x-coordinate is the same as point E's, but its y-coordinate is the opposite of point E's. Locate and label point F. What are the coordinates? How far apart are points E and F? Explain how you arrived at your answer.

A Progression Toward Mastery						
Assessment Task Item			STEP 1 Missing or incorrect answer and little evidence of reasoning or application of mathematics to solve the problem.	STEP 2 Missing or incorrect answer but evidence of some reasoning or application of mathematics to solve the problem.	STEP 3 A correct answer with some evidence of reasoning or application of mathematics to solve the problem, OR an incorrect answer with substantial evidence of solid reasoning or application of mathematics to solve the problem.	STEP 4 A correct answer supported by substantial evidence of solid reasoning or application of mathematics to solve the problem.
1	a	6.NS.C.5 6.NS.C.6a	Student is unable to answer the question. None of the descriptions are correctly represented with an integer although student may make an effort to answer the question.	Student correctly represents only one of the three descriptions with an integer.	Student correctly represents two of the three descriptions with integers.	Student correctly represents all three descriptions with integers: −800, 960, 230.
	b	6.NS.C.5 6.NS.C.6a 6.NS.C.6c	Student does not attempt to locate and label −800 and 800 and provides little or no evidence of reasoning.	Student attempts to locate and label −800 and 800 but makes an error. For example, both integers are not equidistant from 0. Student may or may not correctly identify the relationship as opposites.	Student accurately locates but *does not label* −800 and 800; student correctly identifies the relationship between the integers as opposites. OR Student accurately locates and labels −800 and 800 on the number line but does not identify the relationship between the integers as opposites.	Student accurately locates and labels −800 and 800 on the number line and identifies the relationship between the integers as opposites.

A STORY OF RATIOS

End-of-Module Assessment Task 6•3

	c 6.NS.C.5 6.NS.C.6a	Student response is incorrect, and no evidence of reasoning, such as an explanation or a diagram, is provided.	Student response is incorrect, but student attempts to answer the question with an explanation and/or diagram that demonstrates an understanding of the word *opposite* although it does not address the meaning of "the opposite of the opposite of $800."	Student response correctly states, "Yes, Mr. Kindle's reasoning is correct." But the explanation and/or diagram provided does not completely explain why Mr. Kindle's statement is correct.	Student response correctly states, "Yes, Mr. Kindle's reasoning is correct." The stance is supported with a valid explanation that demonstrates a solid understanding of the fact that the opposite of the opposite of a number is the number itself.
2	**a** 6.NS.C.7b	Student response is missing.	Student provides an incorrect statement but provides some evidence of understanding the ordering of rational numbers in the written work.	Student response provides a correct ordering of -10 and -20 but without units and reference to the context of the situation.	Student response is correct. Student provides the statement: $-10°F$ is warmer than $-20°F$ or $-20°F$ is colder than $-10°F$. OR Student provides some other explanation that contains a valid comparison of the two temperatures.
	b 6.NS.C.7a 6.NS.C.7b	Student response is missing.	Student attempts to write an inequality statement, but the statement is incorrect and does not include all three numbers. OR The incorrect inequality statement lists all three numbers but does not list 10 as the greatest value.	Student writes an inequality statement that orders the three values with 10 as the greatest number, but the statement contains an error. For example, $-10 < -20 < 10$.	The correct answer is given as an inequality statement of $-20 < -10 < 10$ or $10 > -10 > -20$, and 10 degrees is the warmest temperature.
	c 6.NS.C.7c	Student response is missing.	Student response explains how to use a number line to find the number of degrees below zero the temperature is at noon, but the use of absolute value is not included in the explanation, or it is referenced incorrectly, such as $\lvert -10 \rvert = -10$.	Student response includes a correct explanation and understanding of absolute value: $\lvert -10 \rvert = 10$, but the temperature at noon is incorrectly stated as -10 degrees below 0.	Student response includes a correct explanation and understanding of absolute value: $\lvert -10 \rvert = 10$. AND The temperature at noon is correctly stated as 10 degrees below 0.

Module 3: Rational Numbers

193

	d 6.NS.C.7c	Student response is missing. OR Student response is an incomplete statement supported by little or no evidence of reasoning.	Student response is incorrect but shows some evidence of reasoning. However, the explanation does not show that as negative numbers increase, their absolute values decrease. Student explanation may or may not be supported with an accurate vertical number line model.	Student response includes "Yes" along with a valid explanation that indicates that as negative numbers increase, their absolute values decrease. But a vertical number line model is missing or contains an error.	Student response includes "Yes" along with a valid explanation that indicates that as negative numbers increase, their absolute values decrease. The answer is supported with an accurate vertical number line model representing all three temperatures.
3	a 6.NS.C.6a 6.NS.C.6c	Student response is missing. OR There is little or no evidence of understanding in the work shown to determine the correct location and value of point A.	Student incorrectly locates point A (the opposite of point P) on the number line; however, the location of point A indicates some understanding of an integer's opposite.	Student locates the correct point on the number line for the opposite (1, 2, 3, or 4) based on the integer between 0 and -5 ($-1, -2, -3,$ or -4). However, the opposite is not labeled on the number line as point A. OR Student correctly locates and labels point A, the opposite of point P, but point P does not represent an integer between 0 and -5.	A correct answer of the opposite (1, 2, 3, or 4) is given based on correctly choosing an integer between 0 and -5 ($-1, -2, -3,$ or -4) as point P. The opposite is correctly located on the number line and labeled as point A.
	b 6.NS.C.6c 6.NS.C.7a	Student response is missing. OR There is little or no evidence of understanding in the work shown to determine the correct location and value of point B.	Student incorrectly locates point B on the number line; however, the location of point B on the number line indicates that point B is not equal to point P.	Student correctly locates a point on the number line to the left of point P; however, the point is not labeled as point B. OR Student correctly locates and labels point B even though point P does not represent an integer between 0 and -5.	Point B is correctly graphed and labeled on the number line. The point is to the left of point P on the number line; for example, if point P is -3, point B could be -5.

A STORY OF RATIOS

End-of-Module Assessment Task 6•3

	c 6.NS.C.6c 6.NS.C.7a	Student response is missing. OR There is little or no evidence of understanding in the work shown to determine the correct location and value of point C.	Student incorrectly locates point C on the number line; however, the location of point C on the number line indicates that point C is not equal to point P.	Student correctly locates a point on the number line to the right of point P; however, the point is not labeled as point C. OR Student correctly locates and labels point C, even though point P does not represent an integer between 0 and -5.	Point C is correctly graphed and labeled on the number line. The point is to the right of point P on the number line; for example, if point P is -3, point C could be 0.
	d 6.NS.C.6c	Student response is missing. OR There is little or no evidence of understanding in the work shown to determine the correct location and value of point D.	Student incorrectly locates point D on the number line; however, the location of point D is to the right of point P although not halfway between the integer to the right of point P and point P.	Student correctly locates the number that is halfway between point P and the integer to the right of point P; however, the point is not labeled as point D. OR Student correctly locates and labels point D even though point P does not represent an integer between 0 and -5. OR Student locates and labels point D as the number that is halfway between point P and the integer to the *left* of point P.	Student correctly graphs and labels point D on the number line. The point is exactly halfway between point P and the integer to the right of point P on the number line; for example, if point P is -3, point D would be -2.5.
4	a 6.NS.C.5	Student response is missing. OR Student makes an effort to answer the question, but none of the responses are correct.	Student response includes 1, 2, 3, or at most 4 locations represented with correct integers.	Student response includes 5 locations represented with correct integers.	Student response includes all 6 locations represented with the correct integers: 5,343, -210, 614, 50, 0, -435.
	b 6.NS.C.5 6.NS.C.7c 6.NS.C.7d	Student response is missing. OR Student makes an effort to answer the question, but the explanation does not provide any evidence of understanding.	Student attempts to provide an explanation, and the explanation is supported with some evidence of reasoning, but it is incomplete. For example, "Positive and negative numbers tell Julia about sea level."	Student response includes an explanation with evidence of solid reasoning, but the explanation lacks details. For example, "Positive and negative numbers tell Julia how far from sea level a location is."	Student response is correct. An accurate and complete explanation is given, stating that a positive number indicates an elevation above sea level, and a negative number indicates an elevation below sea level.

Module 3:   Rational Numbers

195

	c 6.NS.C.7b 6.NS.C.7c	Student responses are missing, and/or student only partially fills in the chart.	Student fills in the chart attempting to order the elevations and find their absolute values, but more than two numerical errors are made. OR Student fills in the chart and correctly finds the absolute value of each number but does not order the elevations from least to greatest or from greatest to least.	Student fills in the chart ordering the elevations and listing their absolute values, but one or two numbers are incorrect. OR Student fills in the chart and correctly finds the absolute value of each number; however, the elevations are ordered from *greatest to least* rather than least to greatest.	Student response is correct and complete. The chart is accurately completed with elevations ordered from least to greatest and their respective absolute values recorded.
	d 6.NS.C.5 6.NS.C.7c 6.NS.C.7d	Student responses are missing. OR Student circles the row with zeros in the chart to represent sea level but provides no further explanation.	Student circles the row with zeros in the chart to represent sea level and provides an explanation that contains some evidence of reasoning although the explanation may be incomplete or contain inaccurate statements.	Student circles the row with zeros in the chart to represent sea level AND provides a valid explanation, but it lacks details. It is supported with some evidence of reasoning though it may be general in nature. For example, "Elevations below sea level will have different absolute values."	Student circles the row with zeros in the chart to represent sea level, AND an accurate explanation is given and is supported with substantial evidence that sea levels below zero have opposite absolute values as their elevations, and sea levels above zero have the same absolute values as their elevations.
5	a 6.NS.C.8	Student response is missing. OR All 4 points are inaccurately located.	Student accurately locates and labels 1–2 points.	Student accurately locates and labels 3 points.	Student accurately locates and labels all 4 points.
	b 6.NS.C.8	Student response is missing.	Student response is incorrect, AND neither coordinate is stated as a negative number.	Student response is incorrect, but one of the coordinates is correct. For example, $(-6, 3)$ is the response, and the x-coordinate is correct.	Student provides a correct answer expressed as an ordered pair where both the x- and y-coordinates are negative numbers. For example, $(-6, -3)$.

End-of-Module Assessment Task 6•3

	c 6.NS.C.8	Student response is missing. OR An incorrect answer is given with little or no application of mathematics used to solve the problem.	Student provides an incorrect answer for the distance but demonstrates some evidence of understanding how to find the distance between the points although a significant error was made.	Student response correctly states a distance of 19 units, but the work shown does not adequately support the answer. OR An incorrect answer for the distance is given, but the work shown demonstrates a correct process with a minor error. For example, student made an error in the addition or miscounted when using the number line.	Student response is complete and correct. The distance between the points is found to be 19 units, and an accurate and complete explanation, process, and/or diagram is provided to support the answer.
	d 6.NS.C.8	Student response is missing.	Student response is incorrect, and neither coordinate is stated correctly.	Student response is incorrect, but one of the coordinates is correct. For example, $(5, -2)$ is the response, and the x-coordinate is correct.	Student response is correct and complete. Point E's coordinates are $(5, 2)$.
	e 6.NS.C.6b 6.NS.C.8	Student response is missing. OR Student makes an effort to answer the question, but the answer and/or explanation does not provide any evidence of understanding.	Student does not arrive at the correct coordinates for point F and may or may not arrive at the correct distance between points E and F. But there is some evidence of understanding how to locate a point related to point E and/or how to find the distance between the two points.	Student response is partially correct. Point F is correctly located and labeled, and its coordinates are given as $(5, -2)$, but student is unable to arrive at the correct distance between points E and F or is unable to explain the process accurately.	Student correctly completes all 3 tasks. Point F is correctly located and labeled on the coordinate grid, and its coordinates are given as $(5, -2)$. The distance between points E and F is 4 units and is supported with substantial evidence of reasoning.

Module 3: Rational Numbers

A STORY OF RATIOS   End-of-Module Assessment Task   6•3

Name _____   Date _____

1. Mr. Kindle invested some money in the stock market. He tracks his gains and losses using a computer program. Mr. Kindle receives a daily email that updates him on all his transactions from the previous day. This morning, his email read as follows:

   Good morning, Mr. Kindle,

   Yesterday's investment activity included a loss of $800, a gain of $960, and another gain of $230. Log in now to see your current balance.

   a. Write an integer to represent each gain and loss.

Description	Integer Representation
Loss of $800	−800
Gain of $960	960
Gain of $230	230

   b. Mr. Kindle noticed that an error had been made on his account. The "loss of $800" should have been a "gain of $800." Locate and label both points that represent "a loss of $800" and "a gain of $800" on the number line below. Describe the relationship of these two numbers when zero represents no change (gain or loss).

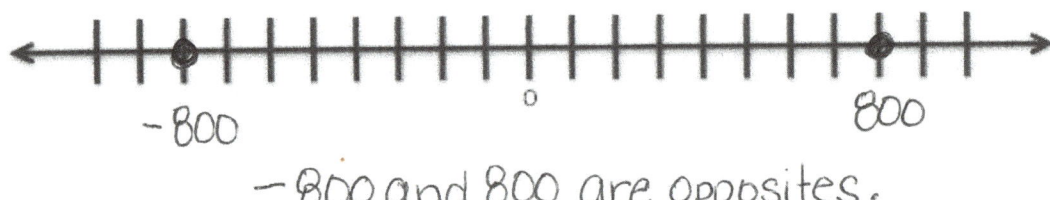

−800 and 800 are opposites.

198   Module 3:   Rational Numbers

c. Mr. Kindle wanted to correct the error, so he entered −(−$800) into the program. He made a note that read, "The opposite of the opposite of $800 is $800." Is his reasoning correct? Explain.

Yes, he is correct. The opposite of 800 is −800, and the opposite of that is 800.

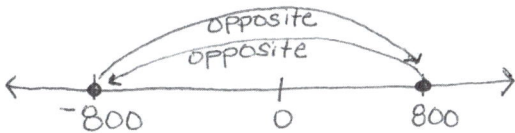

2. At 6:00 a.m., Buffalo, NY, had a temperature of 10°F. At noon, the temperature was −10°F, and at midnight, it was −20°F.

a. Write a statement comparing −10°F and −20°F.

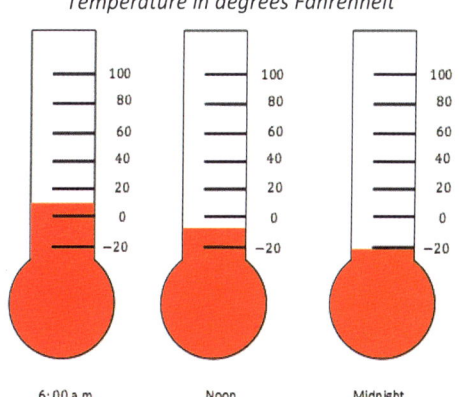

−10°F is warmer than −20°F.

b. Write an inequality statement that shows the relationship between the three recorded temperatures. Which temperature is the warmest?

−20 < −10 < 10
10°F is the warmest temperature.

c. Explain how to use absolute value to find the number of degrees below zero the temperature was at noon.

$|-10| = 10$ The temperature at noon was 10° below zero.

d. In Peekskill, NY, the temperature at 6:00 a.m. was $-12°F$. At noon, the temperature was the exact opposite of Buffalo's temperature at 6:00 a.m. At midnight, a meteorologist recorded the temperature as $-6°F$ in Peekskill. He concluded that "For temperatures below zero, as the temperature increases, the absolute value of the temperature decreases." Is his conclusion valid? Explain and use a vertical number line to support your answer.

$|-12| = 12$ } The absolute values are decreasing.
$|-10| = 10$
$|-6| = 6$

Yes, his conclusion is valid. Absolute value is a number's distance from zero. As the temperature increases from -12 to -10 to -6 they get closer to zero, so their distance from zero is decreasing.

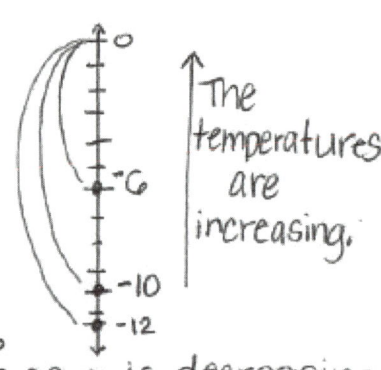
The temperatures are increasing.

3. Choose an integer between 0 and −5 on a number line, and label the point P. Locate and label each of the following points and their values on the number line.

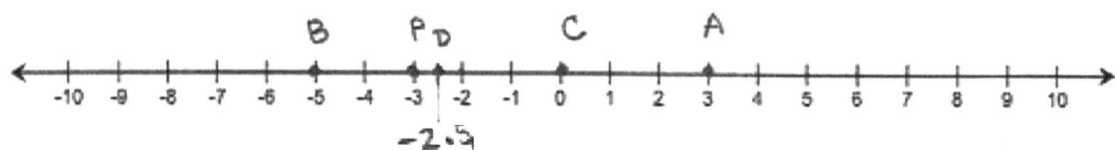

a. Label point A: the opposite of point P.   3

b. Label point B: a number less than point P.   -5

c. Label point C: a number greater than point P.   0

d. Label point D: a number halfway between P and the integer to the right of point P.   -2.5

4. Julia is learning about elevation in math class. She decided to research some facts about New York State to better understand the concept. Here are some facts that she found.

   - Mount Marcy is the highest point in New York State. It is 5,343 feet above sea level.
   - Lake Erie is 210 feet below sea level.
   - The elevation of Niagara Falls, NY, is 614 feet above sea level.
   - The lobby of the Empire State Building is 50 feet above sea level.
   - New York State borders the Atlantic Coast, which is at sea level.
   - The lowest point of Cayuga Lake is 435 feet below sea level.

   a. Write an integer that represents each location in relationship to sea level.

Mount Marcy	5,343
Lake Erie	−210
Niagara Falls, NY	614
Empire State Building	50
Atlantic Coast	0
Cayuga Lake	−435

   b. Explain what negative and positive numbers tell Julia about elevation.

   A negative number means the elevation is below sea level. A positive number means the elevation is above sea level.

c. Order the elevations from least to greatest, and then state their absolute values. Use the chart below to record your work.

Elevations	Absolute Values of Elevations
−435	435
−210	210
(0)	(0)
50	50
614	614
5,343	5,343

d. Circle the row in the table that represents sea level. Describe how the order of the elevations below sea level compares to the order of their absolute values. Describe how the order of the elevations above sea level compares to the order of their absolute values.

The elevations below sea level have absolute values that are their opposites, so the order is opposite. −435 < −210 but 435 > 210.
The elevations above sea level are the same as their absolute values, so the order is the same.
50 < 614 < 5,343

5. For centuries, a mysterious sea serpent has been rumored to live at the bottom of Mysterious Lake. A team of historians used a computer program to plot the last five positions of the sightings.

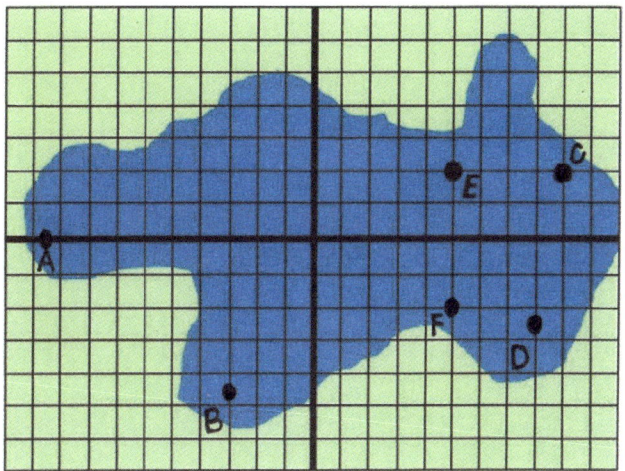

a. Locate and label the locations of the last four sightings: $A\left(-9\frac{1}{2}, 0\right)$, $B(-3, -4.75)$, $C(9, 2)$, and $D(8, -2.5)$.

b. Over time, most of the sightings occurred in Quadrant III. Write the coordinates of a point that lies in Quadrant III.

$(-6, -3)$

c. What is the distance between point A and the point $\left(9\frac{1}{2}, 0\right)$? Show your work to support your answer.

$9\frac{1}{2} + 9\frac{1}{2} = 19$

The distance is 19 units.

d. What are the coordinates of point E on the coordinate plane?

$(5, 2)$

e. Point F is related to point E. Its x-coordinate is the same as point E's, but its y-coordinate is the opposite of point E's. Locate and label point F. What are the coordinates? How far apart are points E and F? Explain how you arrived at your answer.

The coordinates of F are (5, -2). Points E and F are 4 units apart. Since their x-coordinates are the same, I just counted the number of units from 2 to -2 (between their y-coordinates), and that is 4.

This page intentionally left blank